Guide to CMOS Basics, Circuits, & Experiments

by

Howard M. Berlin

Howard W. Sams & Co., Inc.
4300 WEST 62ND ST. INDIANAPOLIS, INDIANA 46268 USA

FIRST EDITION
FIRST PRINTING—1979

International Standard Book Number: 0-672-21654-X
Library of Congress Catalog Card Number: 79-67128

Printed in the United States of America.

Preface

From switches and relays on up to transistors and integrated circuits, digital logic has come a long way. Within the intergated circuit revolution, technology has given us RTL, DTL, TTL, ECL, and CMOS devices, although the majority of the circuit design today is accomplished with either TTL or CMOS. Because of their cost and availability, the user is usually introduced to digital logic with TTL devices. However, because of their low power consumption, ideal for portable battery-powered circuits, CMOS devices are becoming more attractive.

This book was written primarily as a guide for those who have grown up with TTL devices and now are ready to pursue the benefits of CMOS, possibly converting many existing TTL circuits to the lower power CMOS equivalents. For those of you who have not had any experience with digital logic, an excellent course in TTL circuits is presented in a two-volume set entitled *Logic and Memory Experiments Using TTL Integrated Circuits,* published as part of The Blacksburg Continuing Education Series™ by Howard W. Sams & Co., Inc.

In eight chapters, this CMOS book covers what CMOS devices are, their characteristics, design rules, and a series of 22 useful experiments which illustrate many of the concepts discussed. In Chapter 1, the basics of CMOS technology are discussed, covering typical characteristics, numbering systems, design rules, and who makes these devices. Chapter 2 covers the rules and the necessary equipment and components required for the experiments found at the end of each chapter. The remaining chapters discuss the interfacing of CMOS devices with other logic families, transmission gates, analog multiplexers/demultiplexers, monostable and astable multivibrators, counters, the interfacing of LEDs and 7-segment LED displays, and several specialized one-chip system devices.

Although there are many CMOS devices which are either pin-for-pin or functional equivalents of the standard TTL devices, there are also a significant number of CMOS devices which have no TTL equivalent. However, these equivalents are listed in the appendix, along with a number of commercially available useful breadboarding aids.

HOWARD M. BERLIN

This book is dedicated to
Jessica, Judy, and Sara

Contents

CHAPTER 5

CHAPTER 6

CHAPTER 7

CHAPTER 8

APPENDIX A

APPENDIX B

CHAPTER 1

CMOS Basics

INTRODUCTION AND OBJECTIVES

When compared with standard TTL devices, long the mainstay of circuit designers and hobbyists, CMOS devices now present a number of distinct advantages over their older counterparts. This chapter presents an introduction to the basic characteristics that make CMOS devices attractive for the logic circuitry of today. At the completion of this chapter, you will be able to do the following:

- Be familiar with the important characteristics of CMOS devices, especially those which vary with supply voltage.
- Be able to compare these characteristics with standard TTL devices.
- Recognize the design rules that govern the use of CMOS devices.

WHAT IS CMOS?

Complementary Metal Oxide Semiconductor, or CMOS, devices have been called the ideal logic family and offer many advantages over the widely used TTL devices. Before we can discuss the characteristics which make CMOS attractive, we will discuss the basic configuration that allows for the formation of more complex functions.

The most basic of all the CMOS circuits is the inverter, as shown in Fig. 1-1, and consists of a p-channel and an n-channel enhancement mode MOS transistor. The transistors are connected in series across the power supply with the two gates tied together forming the input. The substrate of the p-channel transistor is tied to $+V_{DD}$, while the substrate of the n-channel transistor is tied to $-V_{SS}$, which is usually ground. When the input level to the inverter is at $-V_{SS}$ (logic 0), the substrate of the p-channel transistor is at zero volts relative to the substrate of the n-channel transistor. Consequently, the p-channel transistor is turned on while the n-channel transistor is off. On the other hand, when the input voltage level is at $+V_{DD}$ (logic 1), the reverse is true; the p-channel transistor is off and the n-channel unit is on. In either case, one transistor is always on. In addition, because one of the two transistors is always off, the quiescent power consumption of a complementary transistor pair is extremely low.

The input to the inverter is the equivalent of a 5-pF capacitor, while the resistive component of the input impedance is typically greater than 10^9 ohms.

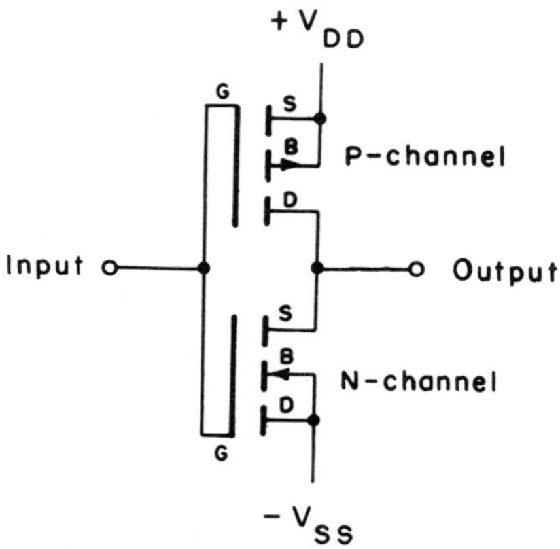

Fig. 1-1. Basic CMOS inverter, formed from p-channel and n-channel transistors.

CMOS SPECIFICATIONS

Unlike TTL devices, CMOS circuits are not restricted to a fixed supply voltage. On the contrary, the supply voltage can range from +3 to +15 volts greater than the negative voltage $-V_{SS}$, which is often at ground potential. For certain series, the supply voltage may be as high as +18 or +20 volts. For most manufacturers, CMOS characteristics are guaranteed for 5-, 10-, and 15-volt operation, and can be easily extrapolated for a particular voltage. On the surface, operation near the lower supply limit of 3 volts may appear attractive; however, the performance of CMOS circuits may be degraded.

By the same token, operation near the upper voltage limit also deserves consideration.

Operation near 3 volts results in an increase in the propagation delay, and output impedance, while the noise immunity is decreased. On the other hand, operation near the 15-volt range results in an increase in power dissipation and the possibility of noise spikes on the power supply, which may exceed the breakdown voltage of the device, typically 20 volts.

Propagation Delay

Compared to standard TTL and low power Schottky TTL, all CMOS devices are basically slow and quite sensitive to capacitive loading. Buffering of outputs will result in lower output impedance which, in turn, reduces the capacitive loading effect. In general, propagation delay is affected by capacitive loading, supply voltage, and temperature.

Both the "on" resistance of the output stage CMOS transistor and the capacitive loading affect the maximum frequency a CMOS device can handle. The "on" resistance is typically a maximum of 1 kΩ so that the maximum frequency can be determined from the equation

$$f_{MAX} = \frac{1}{4.4 R_{ON} C_{LOAD}} \qquad \text{(Eq. 1-1)}$$

Consequently, the maximum frequency is obtained by minimizing the on resistance and the load capacitance. Assuming that the input signal to a CMOS device is perfectly rectangular, then the output waveform can be expressed by the familiar resistor-capacitor charging equation

$$V_o = V_{DD}[1 - e^{-t/R_{ON}C_{LOAD}}] \qquad \text{(Eq. 1-2)}$$

In general, increasing the supply voltage from 5 to 10 volts results in more than a twofold increase in speed of CMOS gates; increasing the supply voltage further to 15 volts results in almost another twofold increase in speed.

Noise Immunity and Noise Margin

Because of the complementary structure of the basic inverter, which is common to all CMOS logic devices, a nearly ideal input-output transfer characteristic results so that the switching point or input threshold is between 45% and 55% of the supply voltage between the logic 0 and logic 1 levels.

Fig. 1-2 illustrates a typical transfer curve which we may use to describe the noise immunity characteristic of CMOS devices. The *noise immunity voltage*, V_{IL} or V_{IH}, is the noise voltage at any one input that goes from input to output. The typical noise immunity is 30% of the supply voltage. Thus, the value for V_{IL} and V_{IH} tells us that the maximum amount of noise that can be added to the input signal when the input signal is within 50 mV of either V_{SS} or V_{DD}.

The difference between the noise immunity voltage and the output voltage V_{OL} or V_{OH} is termed the *noise margin*. The noise margin is the maximum voltage that can be superimposed on the input

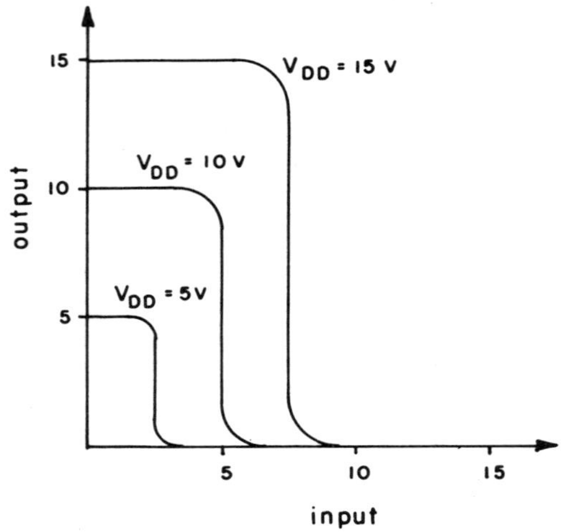

Fig. 1-2. CMOS input/output transfer characteristics.

voltage without changing the logic state or causing the output voltage to exceed the output conditions specified for the V_{IH} and V_{IL} ratings. The use of these characteristics will be explained in Chapter 3 when interfacing CMOS devices with other logic families.

CMOS TYPES

In today's market, CMOS devices are sometimes called different names by various manufacturers. For example, RCA uses the term COS/MOS while Motorola likes the name McMOS. Regardless of the trade names used, there are three major types of CMOS devices in general use: conventional, buffered, and silicon-on-sapphire CMOS. Each type has its pros and cons, but because of their availability and cost, the conventional and buffered types are the most popular.

Conventional CMOS devices were first introduced by RCA and are now commonly referred to as the *A-series*. The CMOS devices within this series have the suffix "A" in conjunction with the code of the device, such as 4000A, 4081A, etc. By tradition, most manufacturers of CMOS devices use the 4000 series numbering system originated by RCA. In addition to the 4000 series, several manufacturers have established their own numbering systems that closely resemble the 4000 series numbers. For example, Motorola's McMOS devices belong to the MC14000 and MC14500 series, while Fairchild has the 34000 series. While none of these numbering systems correspond to the 7400 series used for TTL devices, National Semiconductor, in addition to manufacturing the 4000 series, also has a series of CMOS devices which are pin-for-pin equivalents of the 7400 series of TTL devices, which is the 74C00 series. Depending on the manufacturer, a quad 2-input NAND gate device may have any one of the following numbers: 4011, MC14011, 34011, or 74C00.

Table 1-1. Characteristics of Buffered and Unbuffered Gates*

Characteristic	Buffered Gates	Unbuffered Gates
1. Propagation delay		
$V_{DD} = 5$ V $C_{LOAD} = 50$ pF	150 ns	60 ns
$V_{DD} = 10$ V	65 ns	30 ns
$V_{DD} = 15$ V	50 ns	25 ns
2. Noise immunity	30% of V_{DD} at 5 and 10 V; 27% at 15 V	20% of V_{DD} at 5, 10, and 15 V
3. Noise margin		
$V_{DD} = 5$ V	1 V	0.5 V
$V_{DD} = 10$ V	2 V	1.0 V
$V_{DD} = 15$ V	2.5 V	1.0 V
4. Output impedance		
$V_{DD} = 5$ V, $V_O = \pm 0.4$ V	400 Ω	200 Ω–400 Ω
5. Average input capacitance	1–2 pF	2–3 pF

* Values taken from Table II, RCA Application Note ICAN-6558, *Understanding Buffered and Unbuffered CMOS Characteristics*, by R. E. Funk

Since the introduction of the conventional devices, CMOS devices have been improved primarily by adding a pair of inverters in series to the outputs of conventional devices. These inverters function as buffers, hence the name *buffered* CMOS, which in turn improves the frequency response, output drive, and rise-fall times. Buffered devices are denoted by the suffix "B" in conjunction with the device code, such as 4001B.

The more recent *unbuffered* CMOS devices are those that meet all B-series specifications, except that the outputs are not buffered. In addition, the noise immunity voltages V_{IL} and V_{IH} are specified as 20% and 80% of V_{DD}, respectively, for operation from 5 to 10 volts, and 17% and 83% of V_{DD} for operation from 15 volts. All unbuffered CMOS devices carry the "UB" suffix in conjunction with the device code. Table 1-1 compares the features of the B and UB series.

DESIGN RULES FOR CMOS CIRCUITS

Other than interfacing CMOS devices with other logic families, which are specifically covered in Chapter 3, there are several basic design rules that must be followed in order to guarantee that the CMOS device or system operates correctly. The simple rules that follow are primarily concerned with the inputs of CMOS devices.

1. The input voltage to any CMOS device should *never* exceed the positive power supply voltage V_{DD} or go less than the negative supply voltage V_{SS}, which is frequently ground. In a number of manufacturer's data books, it is often stated that the input voltage may exceed these limits as long as the input current is limited to 10 mA or less. However, this is a poor design practice.
2. *All* unused inputs must be connected either to V_{SS} or V_{DD}. Unlike TTL devices, inputs of CMOS devices cannot be left floating, otherwise the resultant increase in power dissipation may damage the device.
3. Signals should not be applied to the input of any CMOS device while the power supply is *off*.
4. When the inputs to CMOS devices are wired to edge-card connectors, the inputs should be connected to V_{SS} or V_{DD} via 10-kΩ shunt resistors in case the inputs accidentally become disconnected with the power supply on.
5. The input frequency must be less than the maximum allowable frequency. For CMOS devices, the maximum input frequency is a function of the supply voltage. For a 5-volt supply, the maximum input frequency may be 1 MHz for a particular device; for the same device with a 15-volt supply, it may be 5 MHz. However, increasing the supply voltage to achieve higher operating speeds, in turn, increases the power dissipation of the device.
6. The input signals must be "bounce free," or possess fast rise and fall times. The use of push-button switches, etc., must be debounced. In addition, the rise and fall times of clocked logic must be faster than 15 μs, otherwise the power dissipation rating may be exceeded.
7. Because complementary transistors are used, the wired-OR configuration which is possible with certain TTL devices is not possible with CMOS. However, common busing circuits are possible by adding a transmission gate (see Chapter 4) in series

with the output of the device. Consequently three states are permitted; they are logic 0, logic 1, and open-circuit.

MANUFACTURERS

Although there are over a dozen manufacturers of CMOS integrated circuit devices, the ones listed below offer the most extensive series.

Fairchild Semiconductor
464 Ellis Street
Mountain View, CA 94042

Harris Semiconductor
Box 883
Melbourne, FL 32901

Motorola Semiconductor
Box 20912
Phoenix, AZ 85036

National Semiconductor
2900 Semiconductor Drive
Santa Clara, CA 95051

RCA, Solid State Division
Box 3200
Somerville, NJ 08876

Solid State Scientific
Montgomeryville Industrial Park
Montgomeryville, PA 18936

CHAPTER 2

Performing the Experiments

INTRODUCTION

Following this chapter, you will have the opportunity to perform a wide range of experiments with a variety of CMOS devices, interfacing methods, and interesting applications. This chapter discusses the necessary equipment and components that you will need in order to perform the experiments easily and accurately.

RULES FOR SETTING UP EXPERIMENTS

Throughout this book you will be breadboarding various circuits, either using commercial breadboarding aids, or constructing some of the necessary hardware. If you have had experience with any of the *Blacksburg Continuing Education Series*™ texts, these rules will be familiar. Before you set up any experiment, it is recommended that you do the following:

1. Plan your experiment beforehand. Know what types of results you are expected to observe.
2. Disconnect or turn off *all* power and external sources from the breadboard.
3. Clear the breadboard of all wires and components from previous experiments unless instructed otherwise.
4. Check the wired-up circuit against the schematic diagram to make sure that it is correct.
5. When finished, make sure that you disconnect everything *before* you clear the breadboard of wires and components.

FORMAT FOR THE EXPERIMENTS

The instructions for each experiment are presented in the following format:

Purpose

The material under this heading states your purpose for performing the experiment. It is well for you to have this intended purpose in mind as you conduct the experiment.

Pin Configuration of Integrated-Circuit Chips

The pin configurations are given under this heading for all of the integrated-circuit chips used in the experiment.

Schematic Diagram of Circuit

Under this heading is the schematic diagram of the completed circuit that you will wire up in the experiment. You should analyze this diagram in an effort to obtain an understanding of the circuit *before* you proceed further.

Design Basics

Under this heading is a summary of the design equations, when applicable, that apply to the design and operation of the circuit.

Steps

A series of sequential steps describe the detailed instructions for performing portions of the experiments. Questions are also included at appropriate points. Any numerical calculations are performed easily on many of the pocket-type calculators.

HOW MANY EXPERIMENTS DO I PERFORM?

In this text there are many experiments, covering almost all of the circuits and techniques discussed in this book. However, it will not be necessary to perform every experiment. Do as many as you think that are necessary to gain a good feeling for the design and operation of CMOS circuits.

BREADBOARDING

The breadboard is designed to accommodate the many experiments that you will perform in the chapters to follow. The various integrated-circuit devices, resistors, capacitors, and other components, as well as power and signal connections, all tie directly to

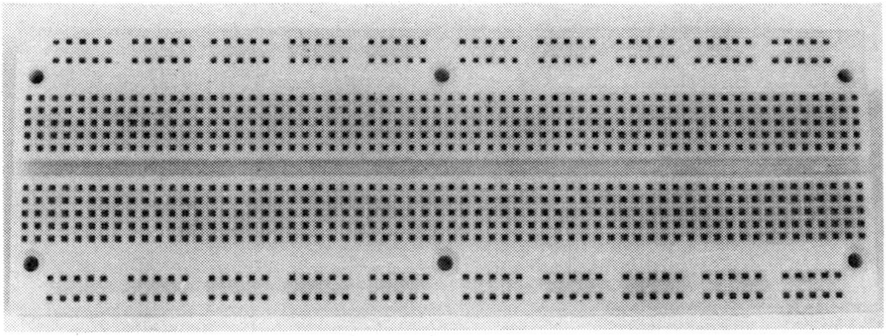

Fig. 2-1. Solderless breadboarding socket used for performing the experiments.

the breadboard. Fig. 2-1 shows the top view of a solderless breadboarding socket, which is manufactured by several companies.

HELPFUL HINTS AND SUGGESTIONS

Tools

Only three tools are really necessary for all of the experiments given in this book:

1. A pair of "long-nosed" pliers
2. A wire stripper/cutter.
3. A small screwdriver.

The pliers are used to

- Straighten out the bent ends of hookup wire that is used to wire the circuits on the breadboard.
- Straighten out or bend the resistor, capacitor, and other component leads to the proper position so that they can be conveniently inserted into the breadboard.

The wire stripper/cutter is used to cut the hookup wire to size and strip about ⅜ inch of insulation from each end.

The screwdriver, if for nothing else, can be used to easily remove the integrated-circuit devices from the solderless breadboarding sockets.

Wire

Only No. 22, No. 24, or No. 26 insulated wire is used, and it must be solid, not stranded!

Breadboarding

- Never insert too large a wire or component lead into a breadboarding terminal.
- Never insert a bent wire. Straighten out the bent end with a pair of pliers before insertion.
- Try to maintain an orderly arrangement of components and wires, keeping all connections as short as possible.

EQUIPMENT

Several pieces of equipment will be required for the experiments.

Oscilloscope

Just about any general-purpose type will do, and it should be a dual trace type.

Frequency Counter

This doesn't have to be an expensive one, but it should have a resolution of 1 Hz for precise measurements. There are several low-cost units available in kit form for less than $80.00 which will perform nicely. Every serious experimenter should have one.

Vom, Vtvm, or Digital Voltmeter

A general-purpose type that is capable of measuring dc voltages is necessary. If you can obtain one, please use a digital type. If a vom is used, its rating should be at least 20 kΩ/V so as not to introduce loading errors.

INPUT/OUTPUT CIRCUITS

In order to perform the experiments smoothly, it will be necessary to have an arsenal of several useful circuits, such as LED monitors, 7-segment displays, debounced switches, etc.

LED Monitors

A *light-emitting diode* (*LED*) *monitor* is a monitor in which the LED is lit for a logic 1 state and unlit for a logic 0 state. A circuit for use with CMOS circuitry is shown in Fig. 2-2, and it is suggested that you have at least 4 LED monitors. When the monitor is used, the schematic symbol of Fig. 2-3 will be shown in the *Schematic Diagram of Circuit* section of the experiment.

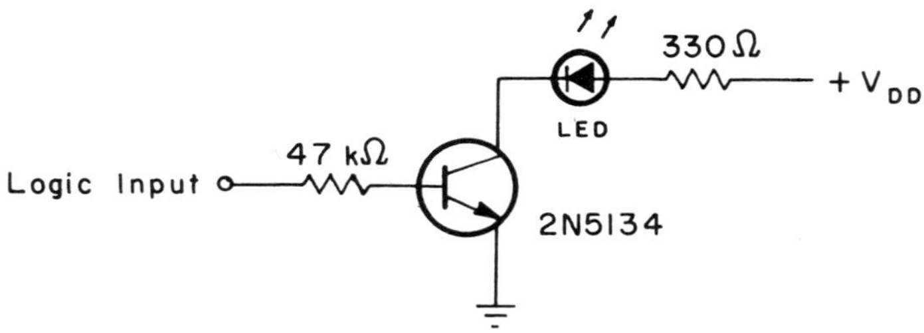

Fig. 2-2. LED monitor.

Fig. 2-3. LED monitor schematic symbol.

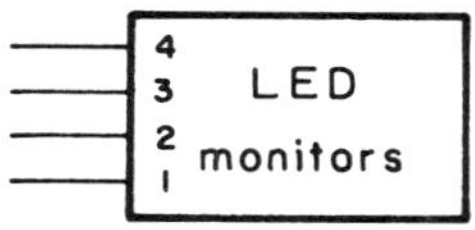

Logic Switches

A *logic switch* is a mechanical spdt switch that applies either a logic 0 or a logic 1 at its output terminal, as shown by the circuit of Fig. 2-4.

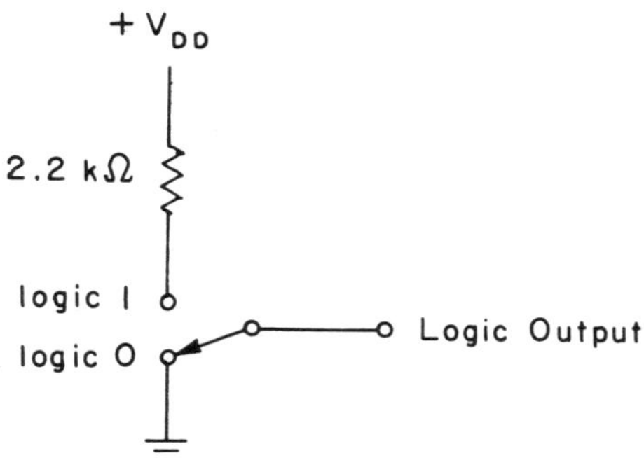

Fig. 2-4. CMOS logic switch.

Usually you will need logic switches in groups of four, primarily representing the inputs of a 4-bit BCD number. Fig. 2-5 shows the schematic diagram that will be used to represent a series of logic switches.

Debounced Switches (Pulsers)

In a *mechanical* logic switch, contact bounce (i.e., the uncontrolled making and breaking of contact when the switch contacts

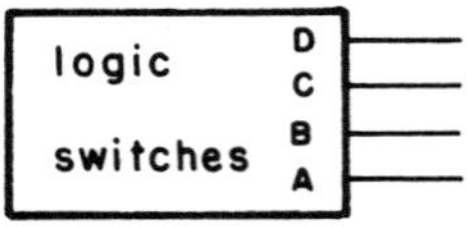

Fig. 2-5. CMOS logic switch schematic symbol.

are opened or closed) is a common occurrence. In most applications, it is extremely important that the output from a switching action be bounce-free, or *debounced.* The circuit shown in Fig. 2-6 is useful as a push-button debounced switch or pulser. Using a

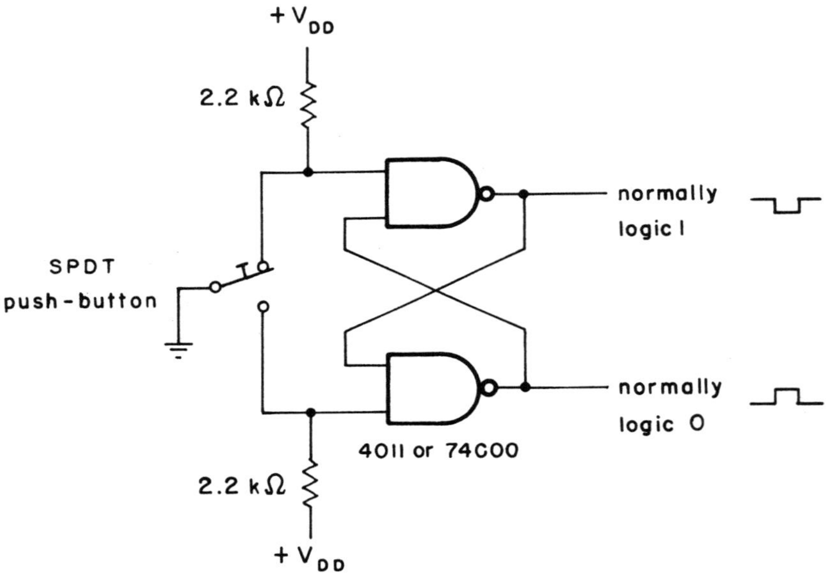

Fig. 2-6. CMOS debounced switch, or pulser.

pair of cross-coupled NAND gates, the pulser has complementary logic 0 and logic 1 outputs. The NAND gate can be either a 74C00 or 4011 device. Fig. 2-7 shows the schematic symbol that will be used to represent a pulser.

Clock

A *clock* is an astable multivibrator that produces a train of square waves alternating between logic 0 and logic 1. A suitable variable

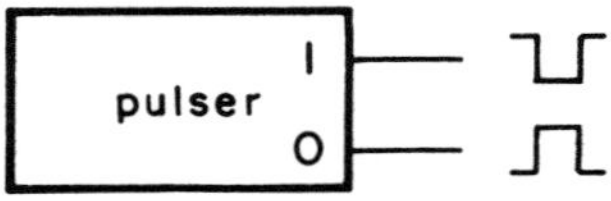

Fig. 2-7. CMOS pulser schematic symbol.

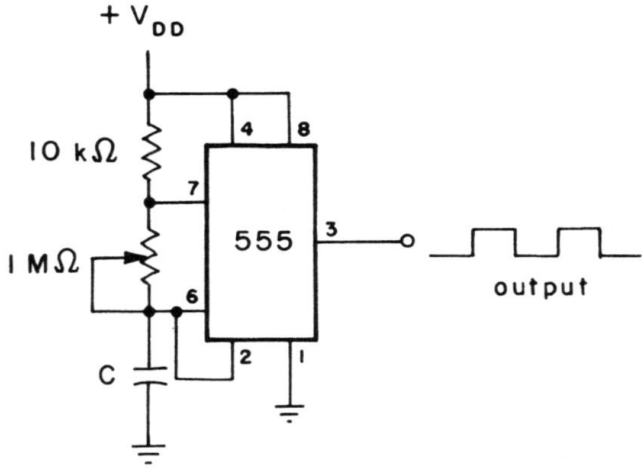

Fig. 2-8. Adjustable frequency clock.

frequency clock for CMOS experiments is shown in Fig. 2-8, built around the 555 timer. A short description of the 555 timer is given in Chapter 5, or, for a more extensive discussion, refer to *The 555 Timer Applications Sourcebook, With Experiments,* published by Howard W. Sams & Co., Inc. Fig. 2-9 shows the schematic symbol that will be used to represent a clock.

Fig. 2-9. Clock schematic symbol.

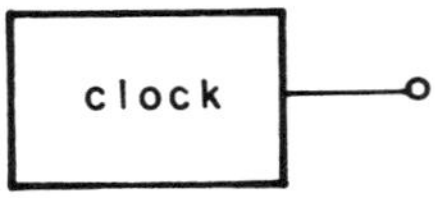

Stable Frequency Reference

A *stable frequency reference* is an astable multivibrator whose output frequency is controlled by a quartz crystal, instead of external capacitors and resistors. As shown in Fig. 2-10, this circuit uses an MK5009 integrated circuit oscillator/divider (MOSTEK Corp.) to generate frequencies from 1 MHz down to 0.01 Hz in 1-decade steps. Programming the MK5009 to generate the various output frequencies is accomplished with a thumbwheel switch to produce the 4-bit BCD input code according to Table 2-1.

The schematic symbol shown in Fig. 2-11 will be used to designate a frequency reference. In addition, the reference frequency will also be indicated (e.g., 1 kHz).

7-Segment LED Display

To easily determine the state of binary counters or 4-bit BCD input data, we can use a *7-segment LED display*. For CMOS circuits,

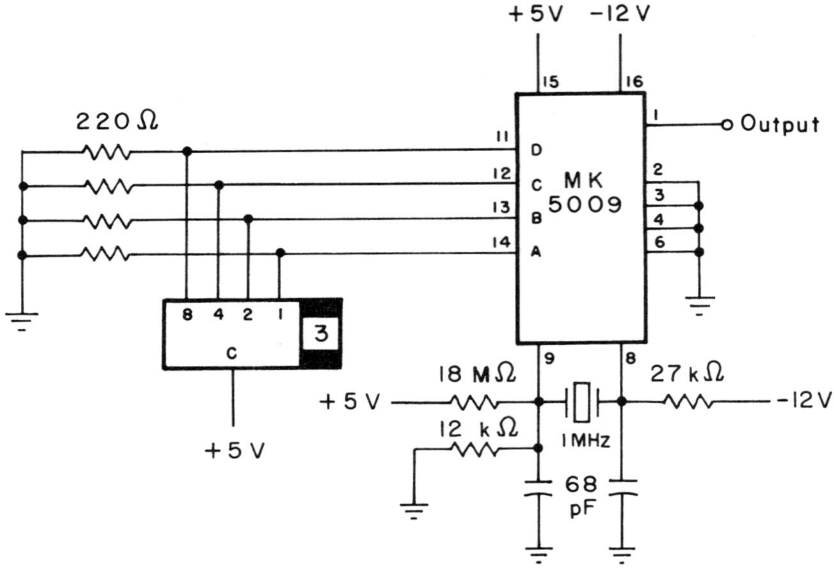

Fig. 2-10. Experimenter-type stable frequency reference.

Table 2-1. Input Code for Programming the MK5009

BCD Inputs D	C	B	A	Divisor	Output Frequency
0	0	0	0	$\div 10^0$	1 MHz
0	0	0	1	$\div 10^1$	100 kHz
0	0	1	0	$\div 10^2$	10 kHz
0	0	1	1	$\div 10^3$	1 kHz
0	1	0	0	$\div 10^4$	100 Hz
0	1	0	1	$\div 10^5$	10 Hz
0	1	1	0	$\div 10^6$	1 Hz
0	1	1	1	$\div 10^7$	0.1 Hz
1	0	0	0	$\div 10^8$	0.01 Hz

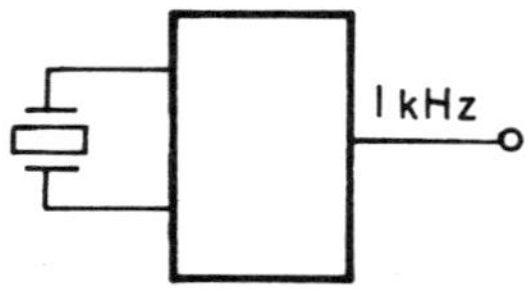

Fig. 2-11. Frequency reference schematic symbol.

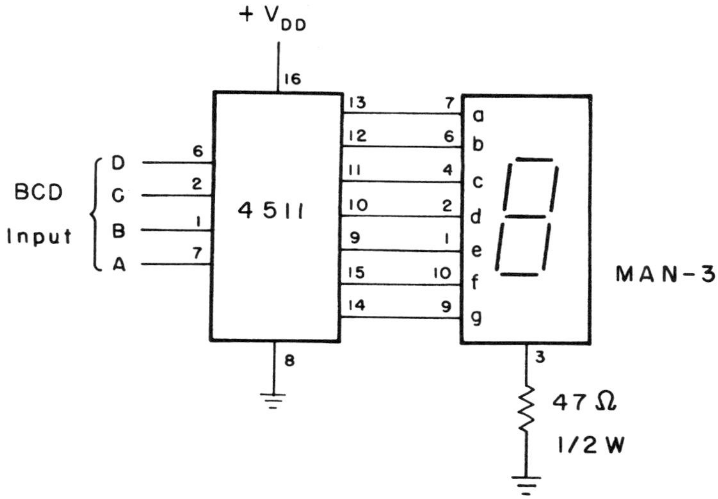

Fig. 2-12. Seven-segment LED display with bcd input.

the circuit shown in Fig. 2-12 uses a 4511 decoder/driver and a *common-cathode* display such as a MAN-3 (0.125 inch display) or a DL704 (0.3 inch) or equivalent. The schematic symbol for a 7-segment LED display is shown in Fig. 2-13. More information on LED displays is presented in Chapter 7.

Fig. 2-13. Schematic symbol for 7-segment LED display.

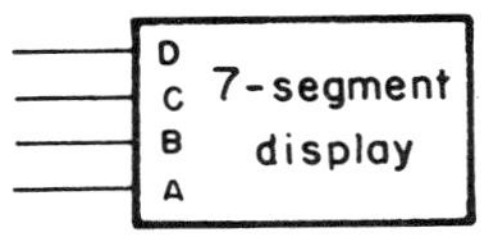

COMPONENTS

A list of all the various components needed to perform all the experiments in this book is shown in Table 2-2.

Table 2-2. Components for Experiments in this Book

Quantity	Description
1	Resistor, 47 ohms, ½ watt
1	Resistor, 150 ohms
1	Resistor, 330 ohms
1	Resistor, 560 ohms
2	Resistor, 1000 ohms
1	Resistor, 4700 ohms
1	Resistor, 10 KΩ

Table 2-2. — Cont.

Quantity	Description
1	Resistor, 12 KΩ
1	Resistor, 15 KΩ
2	Resistor, 24 KΩ
1	Resistor, 27 KΩ
2	Resistor, 39 KΩ
1	Resistor, 47 KΩ
1	Resistor, 56 KΩ
1	Resistor, 100 KΩ
1	Resistor, 220 KΩ
1	Resistor, 330 KΩ
1	Resistor, 390 KΩ
1	Resistor, 620 KΩ
1	Resistor, 1 MΩ
1	Resistor, 10 MΩ
1	Potentiometer, 1MΩ, single turn
1	Capacitor, 0.001 μF
1	Capacitor, 0.01 μF
1	Capacitor, 0.047 μF
2	Capacitor, 0.1 μF
1	Capacitor, 0.33 μF
1	Capacitor, 1 μF
1	Capacitor, 5 μF
1	Capacitor, 10 μF
1	Capacitor, 100 μF
1	IC, 555 timer (8-pin DIP)
1	IC, 7400 quad 2-input NAND gate (TTL)
1	IC, 7473 dual JK flip-flop (TTL)
1	IC, 4001 quad 2-input NOR gate
1	IC, 4016 quad bilateral switch
1	IC, 4017 decade counter
1	IC, 4018 presettable Johnson counter
1	IC, 4022 1-of-8 decoder/counter
1	IC, 4046 phase-locked loop
1	IC, 4047 monostable/astable multivibrator
1	IC, 4049 hex inverting buffer/driver
1	IC, 4050 hex noninverting buffer/driver
1	IC, 4051 1-of-8 analog multiplexer/demultiplexer
1	IC, 4081 quad AND gate
1	IC, 4511 BCD to 7-segment LED display decoder/latch/driver
1	IC, 74C14 or 40106 hex Schmitt trigger
1	IC, 74C192 or 40192 presettable up/down counter
1	IC, HCTR0320 (Hughes) phase detector/programmable divider
2	Diode, 1N914 or equivalent
2	LED, red, green, or yellow
1	7-segment common cathode LED display, or equivalent
1	Transistor, npn, 2N2222 or equivalent

NOTE: Fig. 2-14 shows the type of hardware that you will generally need for the experiments in this book.

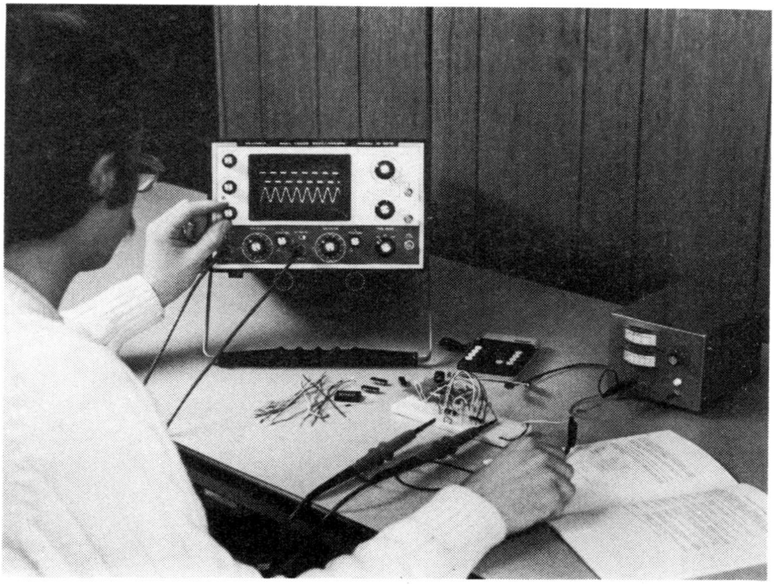

Fig. 2-14. Equipment and hardware used for experiments in this book.

CHAPTER 3

Interfacing CMOS With Other Logic Families

INTRODUCTION AND OBJECTIVES

The interfacing of CMOS devices with other logic families ideally requires that

1. Logic levels be compatible.
2. Both families operate from a single supply voltage.
3. The devices be kept at safe power dissipation levels.
4. The noise immunity be maintained.

In this chapter a number of useful techniques will be discussed that will enable CMOS devices to be interfaced with other logic families and operational amplifiers, especially when the power supply voltages are not the same.

At the completion of this chapter, you will be able to do the following:

- Describe suitable methods for the interfacing of
  - CMOS and standard TTL devices
  - CMOS and other TTL families, such as low-power Schottky
  - CMOS and PMOS devices
  - CMOS and NMOS devices
  - CMOS and operational amplifiers
- Describe the input and output interfacing requirements of CMOS devices.

CMOS INPUT/OUTPUT CONSIDERATIONS

When interfacing CMOS devices with other logic families, careful attention must be paid to the voltage levels that correspond to logic 0 and logic 1. However, the data sheet for a given CMOS device does not directly specify the input voltage requirements. Instead, the *noise immunity* is specified. As illustrated in Fig. 3-1, any

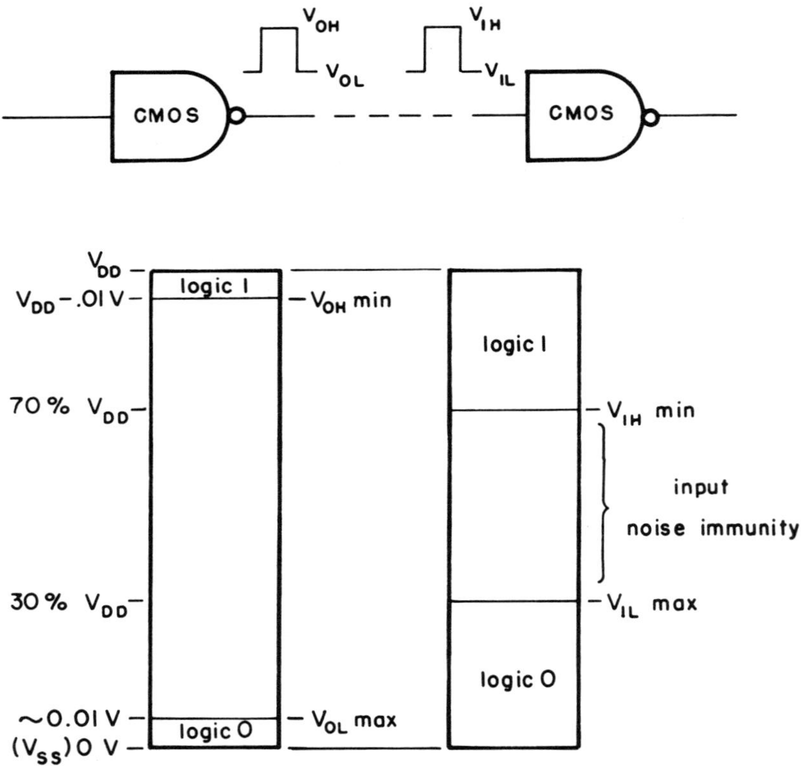

Fig. 3-1. Input and output logic level characteristics of CMOS devices.

input voltage to a CMOS device less than 30% of the supply voltage V_{DD} is at logic 0; any input voltage greater than 70% V_{DD} is at logic 1. Consequently, the input voltage can rise to 30% of V_{DD} or can fall to 70% of V_{DD} without a change of logic level at the output of the device. The noise immunity is then from 30% to 70% of V_{DD}. For a 5-volt supply, this corresponds to levels of 1.5 to 3.5 volts, or a difference of 2.0 volts. Naturally, the noise immunity will increase with increasing supply voltage. At 15 volts, the noise immunity is the difference of 10.5 − 4.5, or 6.0 volts.

The maximum output *sink current* and *source current* ratings vary somewhat with different CMOS devices, and, in addition, vary from manufacturer to manufacturer for the same device. As shown in the equivalent circuit of Fig. 3-2, the CMOS device acts as a *current source* when the output is HIGH, or logic 1. The internal

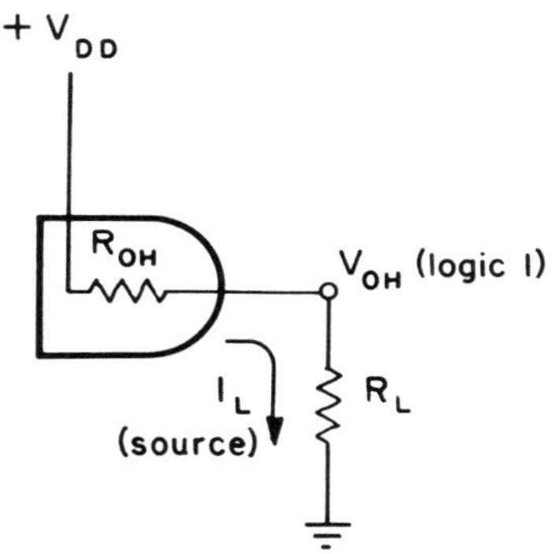

Fig. 3-2. CMOS current source equivalent circuit.

output resistance R_{OH} then causes the output to be less than the supply voltage, according to the following voltage divider relationship:

$$V_{OH} = V_{DD}\left(\frac{R_L}{R_L + R_{OH}}\right) \qquad \text{(Eq. 3-1)}$$

For a logic 1 output, the output resistance is typically in the 200–800 Ω range. When one CMOS device drives another CMOS device, the input impedance is typically 10^{11} ohms. Consequently, the output voltage of the current source is essentially that of the supply. Then for a typical CMOS-to-CMOS connection:

$$I_L\ (\text{source}) = \frac{V_{OH}}{R_L} \qquad \text{(Eq. 3-2)}$$

$$= \frac{V_{DD}}{R_L + R_{OH}}$$

$$= 10\ \text{pA}$$

The maximum source current is typically in the 0.14–0.25 mA range for $V_{DD} = 5$ volts; 0.35–0.6 mA for $V_{DD} = 10$ volts; and 1.0–1.8 mA for $V_{DD} = 15$ volts.

As a current sink, using the equivalent circuit of Fig. 3-3, the output voltage is then LOW, or logic 0. The internal output resistance

R_{OL}, which is essentially that of R_{OH}, causes the output voltage to be equal to

$$V_{OL} = V_{DD}\left(\frac{R_{OL}}{R_L + R_{OL}}\right) \qquad \text{(Eq. 3-3)}$$

while the output sink current is

$$I_L\ (\text{sink}) = \frac{V_{OL}}{R_{OL}} \qquad \text{(Eq. 3-4)}$$

$$= \frac{V_{DD}}{R_L + R_{OL}}$$

$$= 10\ \text{pA}$$

for a CMOS-to-CMOS interface. The maximum sink current is typically 0.4 to 0.8 mA for $V_{DD} = 5$ volts; 0.9 to 1.8 mA for $V_{DD} = 10$

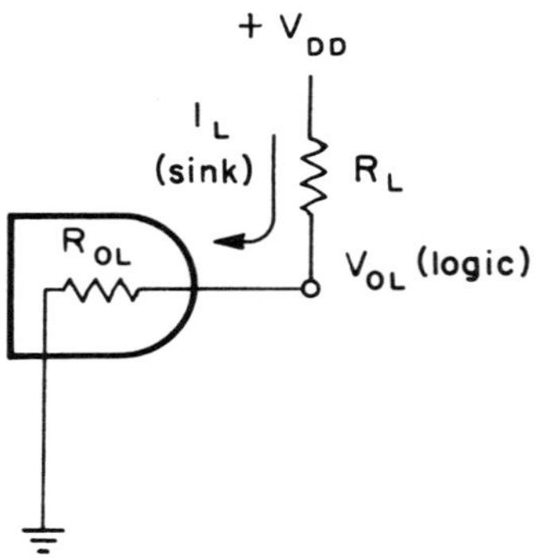

Fig. 3-3. CMOS current sink equivalent circuit.

volts; and 2.0 to 3.6 mA for $V_{DD} = 15$ volts. The dc fan-out is then virtually infinite, but is practically limited to about 50.

CMOS TO TTL INTERFACING

The interfacing of CMOS to TTL requires several compromises. Although both logic families will work at the 5-volt TTL supply voltage, the maximum operating speed of the CMOS device is minimized, being typically 2 MHz vs. 10 MHz and up for TTL. Virtually all CMOS manufacturers imply that their devices are compatible with TTL. This does not mean that the CMOS device can be directly connected to TTL devices, but rather that both CMOS and TTL will operate at the same 5-volt supply voltage. However, the maximum sink current is not sufficient to drive even one standard TTL load.

Another consideration is logic level compatibility, as illustrated in Fig. 3-4. Even if the CMOS device was able to sink the necessary 1.6 mA required by the TTL device for a logic 0, Equation 3-4 with R_{OL} equal to 800 Ω tells us that the output voltage of the CMOS device would be

$$V_{OL} = I_L R_{OL}$$
$$= (1.6 \text{ mA})\ (800\ \Omega)$$
$$= 1.28 \text{ volts}$$

which exceeds the maximum allowable input voltage (V_{IL}) required for a logic 0 for a TTL device.

To overcome the limited output sink current of the CMOS device, the arrangement shown in Fig. 3-5 can be used when interfac-

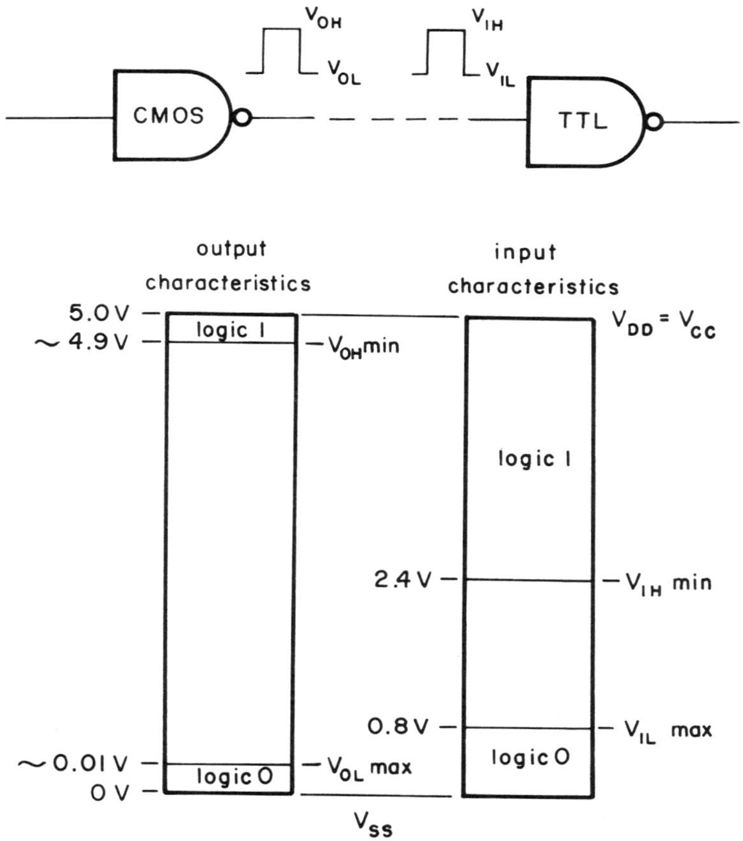

Fig. 3-4. CMOS-to-standard TTL input and output logic level characteristics.

ing CMOS to TTL. The 4049 is an inverting buffer/converter, while the 4050 is noninverting. It should be noted that the CMOS circuitry that is to be interfaced with TTL can be operated with a 3- to 15-volt supply, even though the 4049 or 4050 and the TTL load are operated from a 5-volt supply. The 4049 and 4050 buffers are

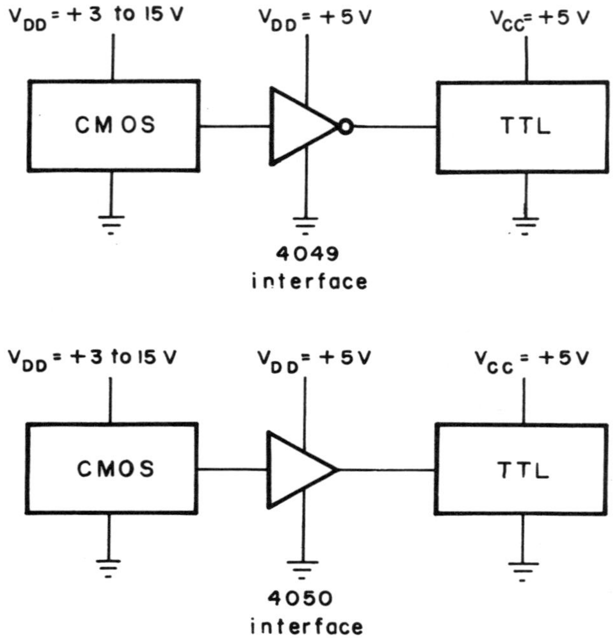

Fig. 3-5. Using a 4049 (inverting) or a 4050 (noninverting) buffer for interfacing CMOS-to-standard TTL devices.

designed with input protection circuitry, thus allowing their input voltage to exceed their supply voltage. Typically this arrangement is capable of driving a maximum of 2 standard TTL loads.

STANDARD TTL TO CMOS INTERFACING

When interfacing standard TTL to CMOS (Fig. 3-6), there are two possibilities:

1. The supply voltage of the CMOS load is the same as the TTL supply (5 volts).
2. The supply voltage of the CMOS load is between 5 and 15 volts.

In either case, the guaranteed minimum active pull-up TTL output voltage of +2.4 volts is lower than the minimum CMOS input voltage required for an acceptable logic 1 input. The +2.4-volt minimum standard TTL output level is often specified at a load current of approximately 400 μA. Since virtually no current (actually 10

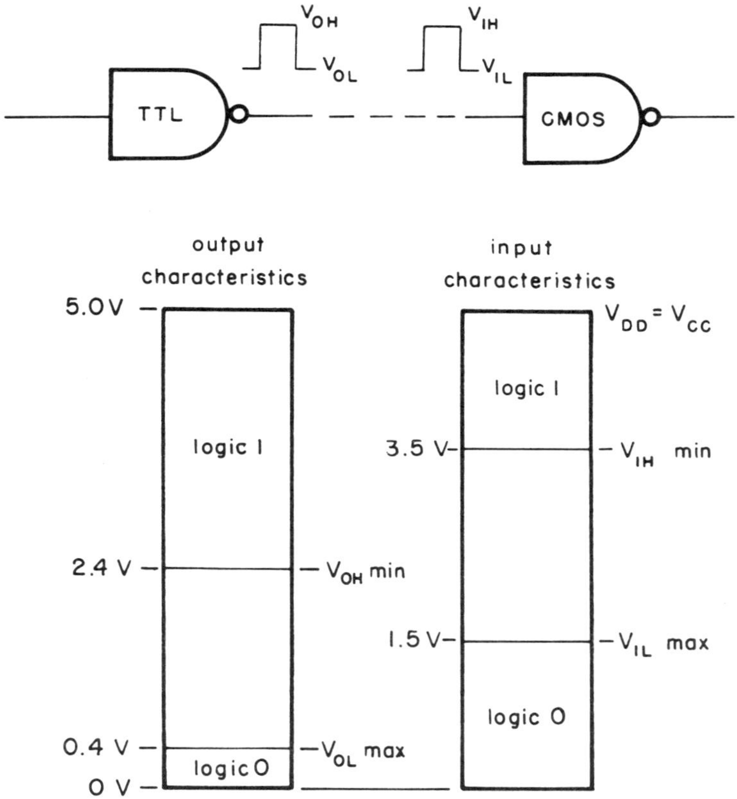

Fig. 3-6. Standard TTL-to-CMOS input and output logic level characteristics.

pA) is drawn by the CMOS device, the minimum TTL logic 1 output voltage is typically 3.4 to 3.6 volts. Consequently, the noise immunity is essentially nil.

To overcome the incompatible logic levels, an external *pull-up* resistor, as shown in Fig. 3-7 is used. To compute the value for the pull-up resistor, several factors must be known or assumed:

1. Fan-out (number of CMOS loads), M
2. Maximum logic 0 collector current of TTL device, I_{OL} max

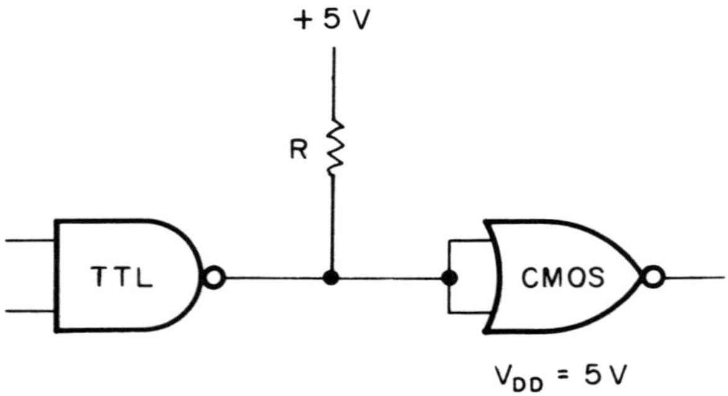

Fig. 3-7. Using a pull-up resistor for interfacing standard TTL to CMOS with equal supply voltages.

3. Maximum logic 1 collector-to-emitter leakage current of TTL device, I_{CE} max
4. The supply voltage, V_{CC} (for TTL) and V_{DD} (for CMOS)
5. Logic 0 and 1 input currents of CMOS device, I_{IL} and I_{IH}
6. Number of TTL driver stages, N

The minimum and maximum values for R, based on Fig. 3-8, are given by:

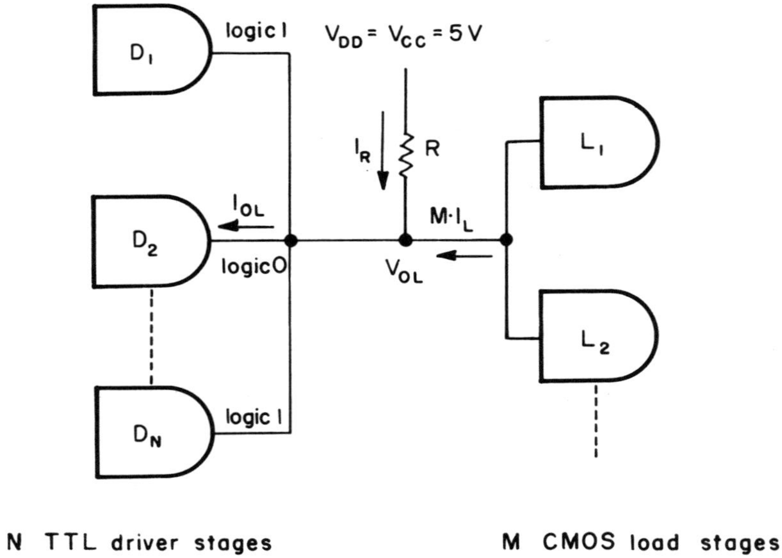

Fig. 3-8. Node currents in standard TTL-to-CMOS interfacing.

$$R_{min} = \frac{V_{DD} - V_{OL}max}{I_{OL} - MI_{IL}} \qquad \text{(Eq. 3-5)}$$

and

$$R_{max} = \frac{V_{CC}min - V_{IH}min}{NI_{CE}max + MI_{IH}} \qquad \text{(Eq. 3-6)}$$

Since the CMOS currents I_{IL} and I_{IH} are both approximately 10 pA and are negligible, the above two equations can be simplified to

$$R_{min} = \frac{V_{DD} - V_{OL}max}{I_{OL}max} \qquad \text{(Eq. 3-7)}$$

and

$$R_{max} = \frac{V_{CC}min - V_{IH}min}{NI_{CE}max} \qquad \text{(Eq. 3-8)}$$

To illustrate the use of Equations 3-7 and 3-8, consider the interfacing of a single TTL device with a single CMOS device, as was shown in Fig. 3-7. Typically, the maximum allowable TTL logic 0 collector current I_{OL} maximum is 1.6 mA, and the TTL logic 1 collector-to-emitter leakage current I_{CE} maximum is 100 μA. Therefore,

$$R_{min} = \frac{(5 - 0.4)V}{16 \text{ mA}} = 288 \ \Omega$$

and

$$R_{max} = \frac{(4.75 - 3.5)V}{(1)(100 \ \mu A)} = 12.5 \text{ k}\Omega$$

TTL open-collector devices, such as the 7401, 7403, 7405, 7406, and 7407, also require an external pull-up resistor when driving CMOS loads. However, the open-collector driver should not be

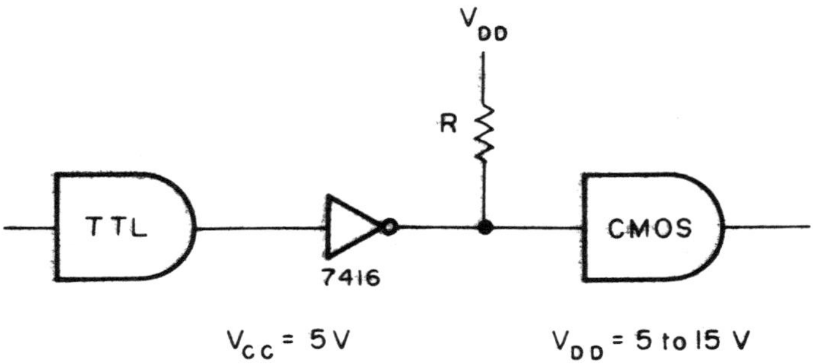

Fig. 3-9. TTL-to-CMOS interface when the CMOS supply voltage is greater than the TTL supply voltage.

connected to other TTL devices, but may be connected to additional CMOS devices.

Since the speed and noise immunity of CMOS devices increases with the supply voltage, a supply voltage that is greater than the 5-volt TTL supply is frequently used. In this case for supply voltages between 5 and 15 volts, open-collector devices, such as the

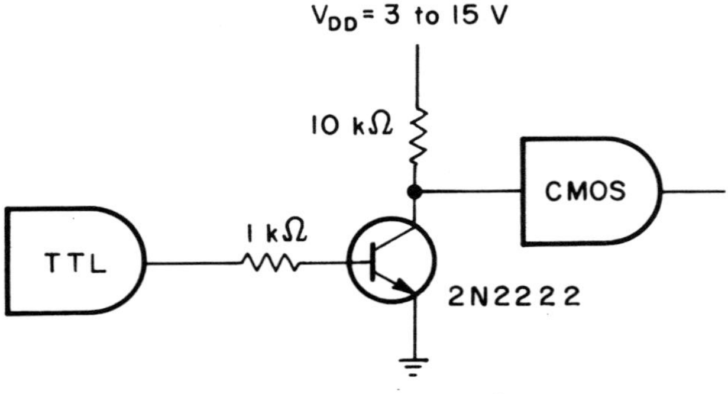

Fig. 3-10. TTL-to-CMOS interface using an npn transistor.

7416, 7417, or 7426, are used as shown in Fig. 3-9. Then the range of the pull-up resistor can be determined from Equations 3-7 and 3-8. Fig. 3-10 shows a TTL-to-CMOS interface using an npn transistor.

OTHER TTL FAMILIES

In addition to *standard* TTL circuits, identified by the "commercial" grade prefix 74, there are other TTL families such as *low-power* (74L), *Schottky* (74S), *low-power Schottky* (74LS), *high-power* (74H) TTL, which trade off power consumption and speed. As of this writing, the low-power Schottky devices have essentially replaced the low-power and high-power TTL devices in most applications not using standard TTL or CMOS circuits.

Since the 74S or 74LS series have markedly lower input current requirements than standard TTL devices, a CMOS device is able to drive a low-power or low-power Schottky device directly. As shown in Fig. 3-11, the low-power device nominally requires a sink current of 0.18 mA for a logic 0 input. Then for an output resistance of the CMOS device equal to 800 Ω, Equation 3-4 shows that

$$\begin{aligned} V_{OL} &= R_{OL}I_L(\text{sink}) \\ &= (800\ \Omega)(0.18\ \text{mA}) \\ &= 0.144\ \text{volts} \end{aligned}$$

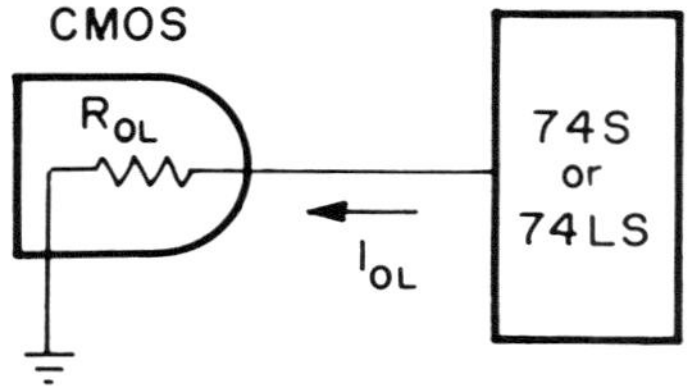

Fig. 3-11. CMOS-to-low power Schottky TTL interface.

which is an acceptable logic 0 input to the low-power device. In fact, most CMOS devices can adequately drive two low-power TTL devices.

When interfacing a low-power or low-power Schottky TTL device to drive a CMOS device, the pull-up resistor method (Fig. 3-7) is commonly used. The resistance is typically between 820 Ω and 10 kΩ.

CMOS-NMOS INTERFACING

N-channel MOS devices, such as random access memories (RAMs), can be operated at the same positive voltage levels as CMOS devices and the NMOS inputs are CMOS compatible. As an example, Fig. 3-12 illustrates the interfacing of CMOS with a 2102 NMOS 1K static RAM. The data output (DO) requires only a single pull-up resistor to assure the proper logic 1 level to the CMOS device. The address lines, data input (DI), read/write (R/W), and chip enable $(\overline{CE})$ signals require no interfacing.

On the other hand, there are other NMOS devices, such as the 2107A 4K dynamic RAM, which use positive and negative supply

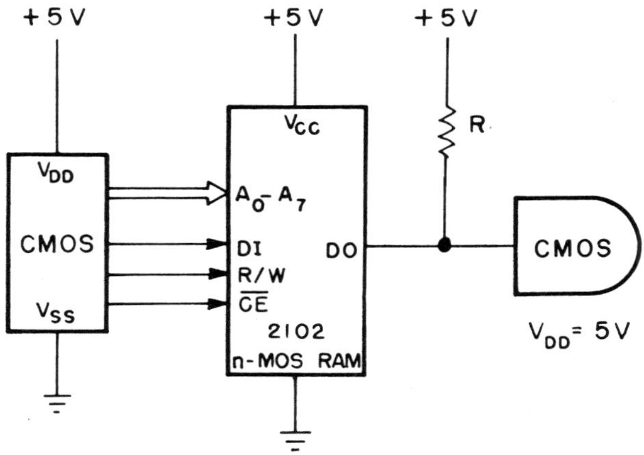

Fig. 3-12. Interfacing CMOS with a 2102 NMOS RAM.

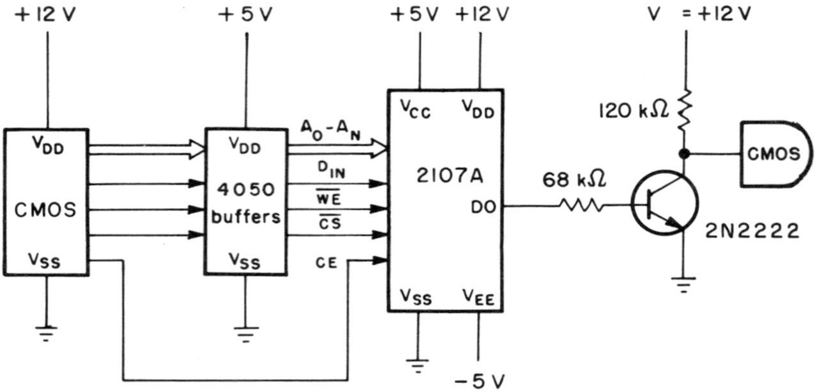

Fig. 3-13. Interfacing CMOS with a 2107A NMOS RAM having two positive supply voltages and a single negative supply voltage.

voltages. As shown in Fig. 3-13, the input signals to the 2107A are translated from +12 to +5 volts with 4050 CMOS buffers. The data output is interfaced back to CMOS circuitry via a transistor buffer connected to $+V_{DD}$.

CMOS-PMOS INTERFACING

P-channel MOS devices generally operate with a single positive ($+V_{CC}$) and a single negative ($-V_{GG}$) supply voltage. As with NMOS devices, PMOS inputs are directly compatible with CMOS

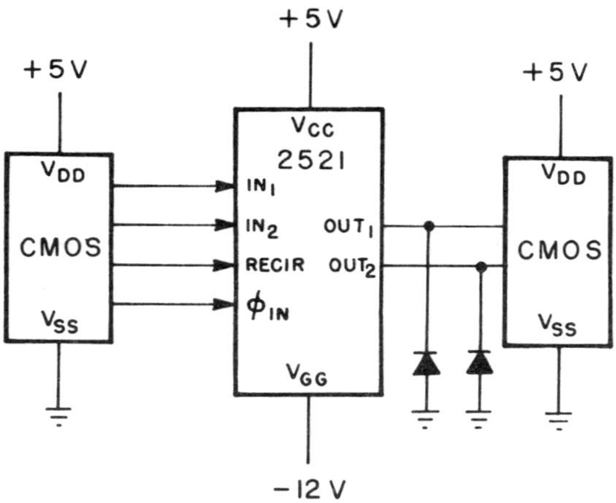

Fig. 3-14. Interfacing CMOS with a 2521 PMOS static shift register having a single positive and negative supply.

outputs. Using a 2521 128-bit static shift register, as illustrated in Fig. 3-14, the only additional component required is a clamp diode from the data output to ground, since the output from the PMOS device will go negative in the LOW output state.

CMOS-OPERATIONAL AMPLIFIER INTERFACING

The output of split-supply op-amp circuits are easily interfaced with CMOS devices, as illustrated by the op-amp comparator circuit of Fig. 3-15, assuming that the CMOS supply is less than the op-amp supply. The two clamp diodes prevent the CMOS input from exceeding the $-V_{SS}$ (or ground) to $+V_{DD}$ range, while the 12 kΩ series resistor limits the output current of the op-amp in case the op-amp output goes toward the negative supply ($-V_{CC}$).

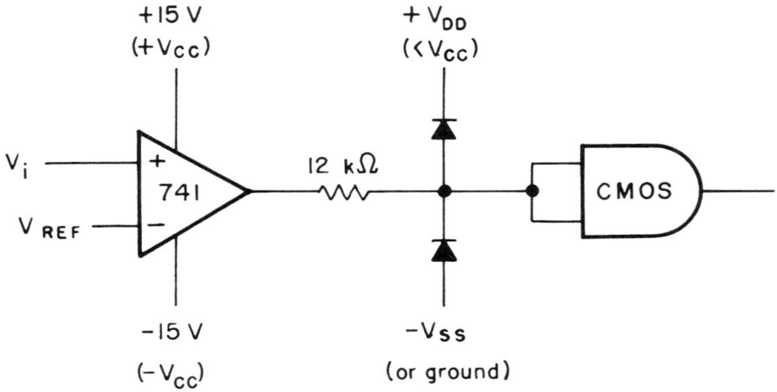

Fig. 3-15. Interfacing an operational amplifier comparator with CMOS.

AN INTRODUCTION TO THE EXPERIMENTS

The three experiments that follow are intended to demonstrate how CMOS devices are easily interfaced with other logic, such as TTL devices. The experiments that you will perform can be summarized as follows:

Experiment No.	*Purpose*
1	Demonstrates how a standard TTL load operating from a 5-volt supply is interfaced with a CMOS circuit with a supply voltage ranging from 3 to 15 volts.
2	Demonstrates the interfacing of a CMOS load to a standard TTL device using a pull-up re-

sistor, with both devices powered from the same 5-volt supply voltage.

3 Demonstrates the interfacing of a CMOS load to a standard TTL device using an npn transistor. The CMOS load is capable of operating over a 3- to 15-volt supply range.

EXPERIMENT 1

Purpose

The purpose of this experiment is to demonstrate the interfacing of a standard TTL load, using a +5-volt supply, to a CMOS circuit with a 3- to 15-volt supply. A 4050 buffer serves as the interface.

Pin Configuration of Integrated Circuit Chips (Fig. 3-16)

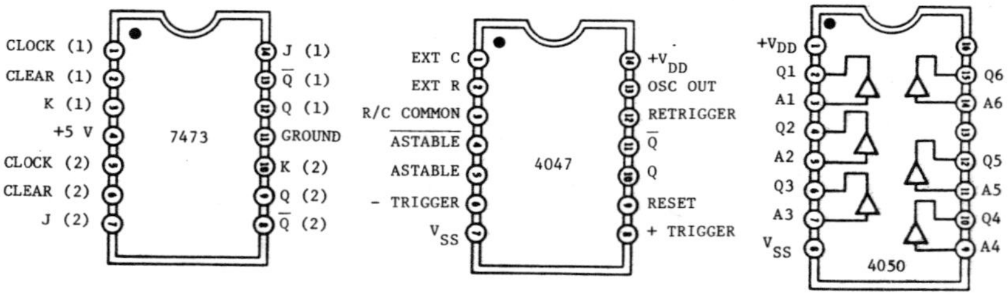

Fig. 3-16.

Step 1

Wire the circuit as shown in the schematic diagram. The 4047 device is wired as an astable multivibrator, and serves as a representative CMOS circuit to be interfaced with a TTL load. Since only one of the six buffers of the 4050 device is used, connect the unused inputs (pins 5, 7, 9, 11, and 14) to the 5-volt TTL supply voltage, since unused CMOS inputs cannot be left floating. The TTL load is simply a 7473 JK flip-flop. The LED and 150 Ω resistor serves as a simple TTL LED monitor.

Step 2

Connect the power sources to the breadboard. Adjust the supply voltage connected to the 4047 integrated circuit at any voltage between 3 and 15 volts. The 4050 buffer is designed with input protection circuitry, allowing its input voltage to exceed its supply voltage.

After the power is connected to the breadboard, the *CMOS* LED monitor at pin 10 of the 4047 should be flashing approximately

Schematic Diagram of Circuit (Fig. 3-17)

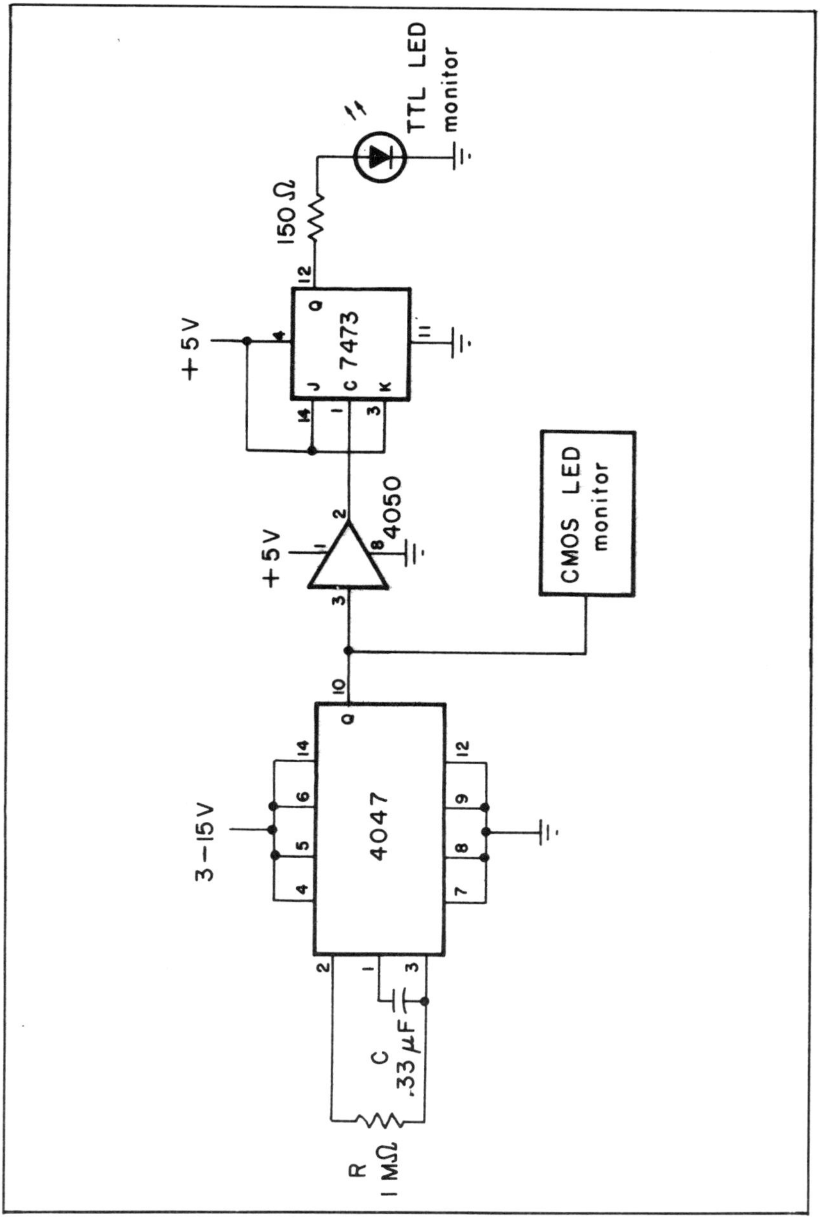

Fig. 3-17.

every second. In addition, the *TTL* LED monitor connected to the output of the 7473 should be flashing once for every two times the CMOS LED monitor is lit. The 7473 simply acts as a divide-by-2 counter.

Step 3

Change the supply voltage connected to the 4047 device to 15 volts. You should notice that the circuit operates exactly as in Step 2.

Step 4

Decrease the supply voltage from 15 volts to 3 volts. Is there any difference in the operation of the circuit?

No. The 4050 CMOS buffer, although itself connected to a 5-volt supply, allows CMOS circuits with any supply voltage between 3 and 15 volts to be connected to a TTL load.

EXPERIMENT 2

Purpose

The purpose of this experiment is to demonstrate the interfacing of a CMOS load to a TTL device using a pull-up resistor. Both the CMOS and TTL devices are powered from a 5-volt supply.

Pin Configuration of Integrated Circuit Chips (Fig. 3-18)

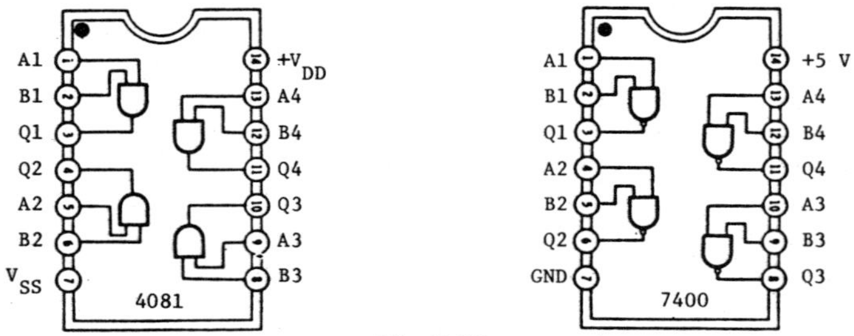

Fig. 3-18.

Design Basics

$$\text{Pull-up Resistor: } R_{min} = 270\ \Omega$$

$$R_{max} = \frac{1.25}{N(0.1\ \text{mA})}$$

where,

N is the number of TTL driver stages.

Schematic Diagram of Circuit (Fig. 3-19)

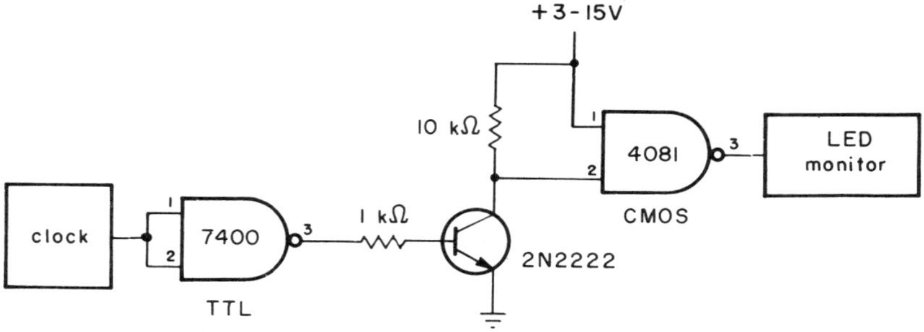

Fig. 3-19.

Step 1

Wire the circuit as shown in the schematic diagram. The clock and the 7400 NAND gate make up the TTL driver circuit while the 4081 AND gate is the CMOS load. Connect the unused inputs of the 4081 (pins 5, 6, 8, 9, 12, and 13) to the supply voltage.

Step 2

Apply power to the breadboard and adjust the frequency of the clock so that LED monitor No. 1 flashes approximately once each second. Since one input of the 2-input CMOS AND gate is tied to logic 1, the output of the AND gate will be the same as the output of the TTL NAND gate. The pull-up resistor is used to overcome the incompatible logic levels that exist between the CMOS and TTL families.

EXPERIMENT 3

Purpose

The purpose of this experiment is to demonstrate the interfacing of a CMOS load to a standard TTL device using an npn transistor. Although the TTL device operates from a 5-volt supply voltage, the CMOS load can be powered from supply voltages over the 3- to 15-volt range.

Step 1

Wire the circuit as shown in the schematic diagram. The clock and the 7400 NAND gate, as in the previous experiment, make up the TTL driver circuit, while the 4081 AND gate is the CMOS load. Connect the unused inputs of the 4081 (pins 5, 6, 8, 9, 12, and 13) to the +3-volt supply.

Pin Configuration of Integrated Circuit Chips (Fig. 3-20)

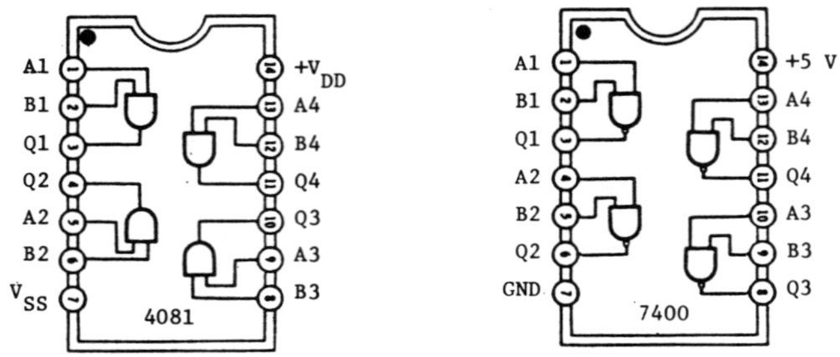

Fig. 3-20.

Schematic Diagram of Circuit (Fig. 3-21)

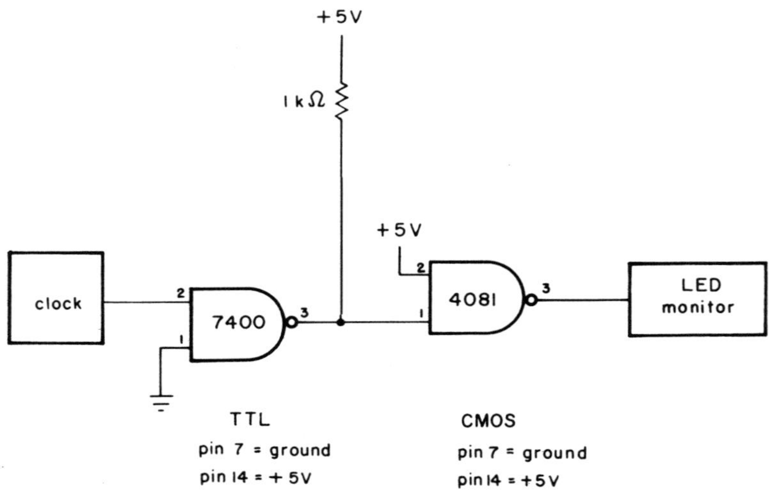

Fig. 3-21.

Step 2

Apply power to the breadboard and adjust the frequency of the clock so that the LED monitor flashes approximately once each second. The npn transistor is used to overcome the incompatible logic levels that exist between the TTL driver (0/5 volts) and the CMOS load (0/3 volts). When used, the transistor stage inverts the input signal so that when the input to the transistor stage is at +5 volts (TTL logic 1), the collector voltage is essentially 0 volts since the transistor is driven into saturation. When the input is zero (TTL logic 0), the collector voltage is nearly that of the CMOS supply, or 3 volts, since the transistor is cut off.

Step 3

Increase the supply voltage to the transistor and the CMOS circuitry to 10 volts. The LED monitor should still be flashing since the transistor circuit allows the CMOS load to properly operate from a supply voltage that is different than the standard +5-supply used for TTL devices.

CHAPTER 4

Transmission Gates and Analog Multiplexers/Demultiplexers

INTRODUCTION AND OBJECTIVES

CMOS transmission gates are unique devices in that they have no TTL equivalent. They have many advantages and can be used for the transmission of analog *or* digital signals, control of operational amplifier circuits, and the simulation of simple switch and logic functions to name a few. Analog multiplexers/demultiplexers are an extension of the basic transmission gate.

At the completion of this chapter, you will be able to do the following:

- Describe the basic components and functions of a transmission gate.
- Describe how transmission gates can be wired to simulate spst, spdt, dpst, and dpdt switches.
- Describe how transmission gates can be wired to simulate an AND gate, an OR gate, JK, and D-type flip-flops.
- Use transmission gates to control the operation of operational amplifier circuits.
- Describe the operation of the 4051, 4052, 4053, 4067, and 4097 multiplexer/demultiplexers.

THE TRANSMISSION GATE

The CMOS transmission gate, sometimes called a bilateral or analog switch, is a unique device in that it has no TTL equivalent.

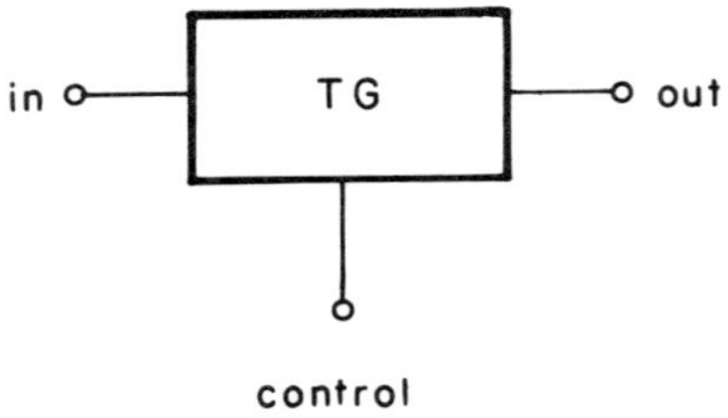

(A) Schematic symbol.

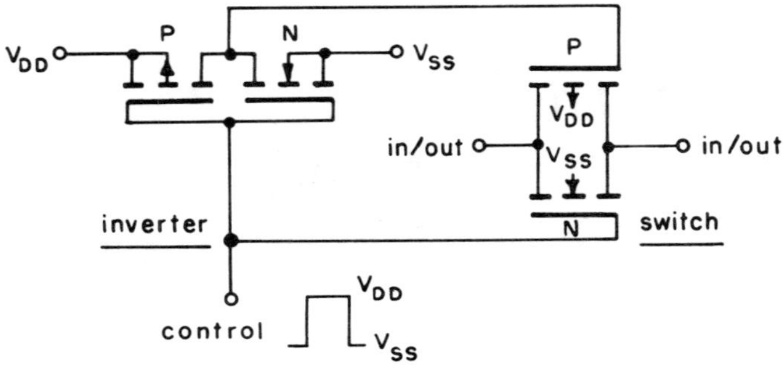

(B) Equivalent circuit.

Fig. 4-1. The transmission gate.

As simple as it is, it is a basic building block in the formation of CMOS flip-flops, shift registers, and counters. Low-level analog and digital signals can be switched to other circuits at rates approaching 10 MHz. Unlike a mechanical switch or relay, the transmission gate exhibits no contact bounce during switching. In this section, the operation and application of transmission gates will be discussed.

Shown in Fig. 4-1B, the transmission gate is composed of 2 pairs of p-channel and n-channel MOS transistors. One pair is connected as a simple inverter while the other pair is connected in parallel and functions as the switch. When the control voltage V_C equals a logic 1 (V_{DD}), the p- and n-channel transistors are switched on, resulting in a low resistance path from input to output. If the control voltage is set at logic 0 (V_{SS}), the transistors are now switched off, resulting in virtually an open circuit from input to output. Since the pair of p- and n-channel transistors that form the switch are connected in parallel, *the input and output terminals are interchangeable,* so that the transmission gate is *bilateral. Unlike any other logic function, digital or analog data can pass through a transmission gate in either direction.*

CMOS transmission gates are made by a number of manufacturers, such as RCA, Intersil, Siliconix, National Semiconductor,

Harris, Fairchild, Motorola, and Solid State Scientific. Perhaps the best known are the 4016 and 4066 quad bilateral switches, as shown by the pin diagram of Fig. 4-2. When the control voltage is at logic 1, the on resistance of the 4016 is rated at 300 Ω, while the on resistance is typically 80 Ω for the 4066. For a logic 0 control voltage, the off resistance for both devices is typically 10^{12} Ω. Consequently, the switching source will not be subjected to loading effects when the transmission gate is off. The range of the input signals must not exceed the power supply limits. For digital signals, the input can vary from ground to +15 volts. For analog signals, the input can vary from −7.5 to +7.5 volts. In either case, the maximum voltage difference $V_{DD} - V_{SS}$ is 15 volts.

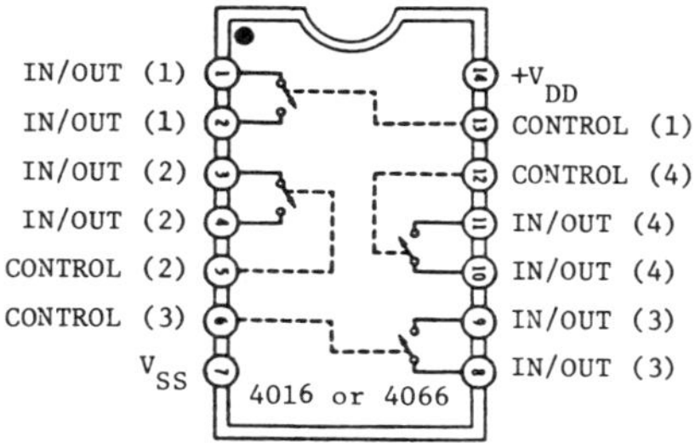

Fig. 4-2. Pin diagram of the 4016/4066 quad transmission gate.

Switch Functions

Either a 4016 or 4066 can be connected to perform a variety of switch functions, as shown in Fig. 4-3. The single-throw functions (spst, dpst) require no additional logic elements, while the double-throw functions require a single inverter, such as a 4069.

As an alternative to the 4016 and 4066 transmission gates, a 4007 combination dual complementary pair plus inverter chip (Fig. 4-4) can be wired as a functional equivalent of a single transmission gate. As shown in Fig. 4-5, the 4007 is wired to act as an spst switch, while the connections shown in Fig. 4-6 represent an spdt switch.

Logic Functions

Given enough NAND gates, it is possible to construct virtually every type of logic function. The same is almost true of transmission gates. Fig. 4-7 illustrates the connections required for use as a digital OR gate. Consequently, the output will be a logic 1 if any of the inputs are at logic 1. These circuits show that transmission gates may be used in a "wired-OR" configuration, which is not possible with other CMOS devices.

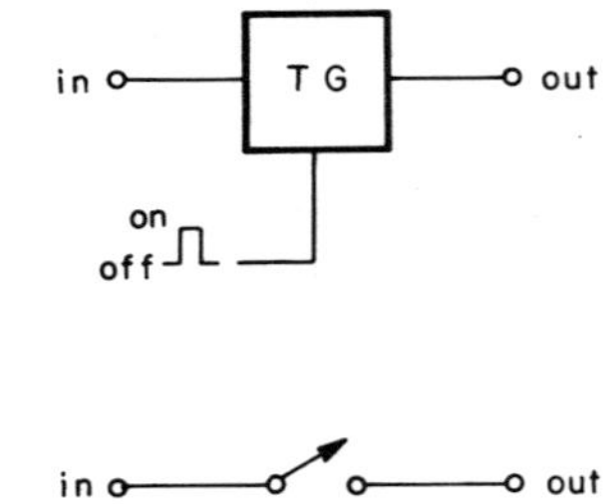

(A) Spst switch.

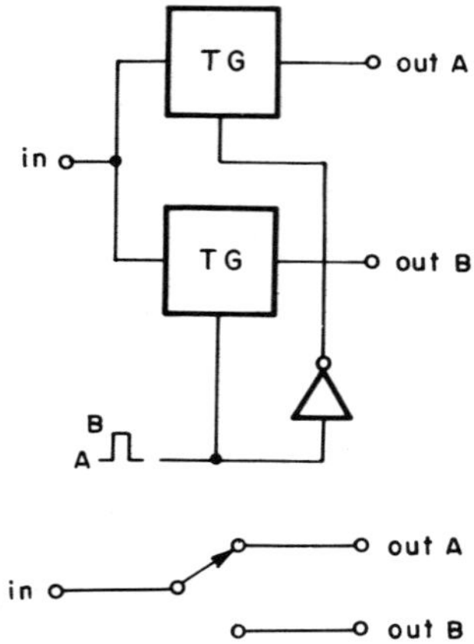

(B) Spdt switch.

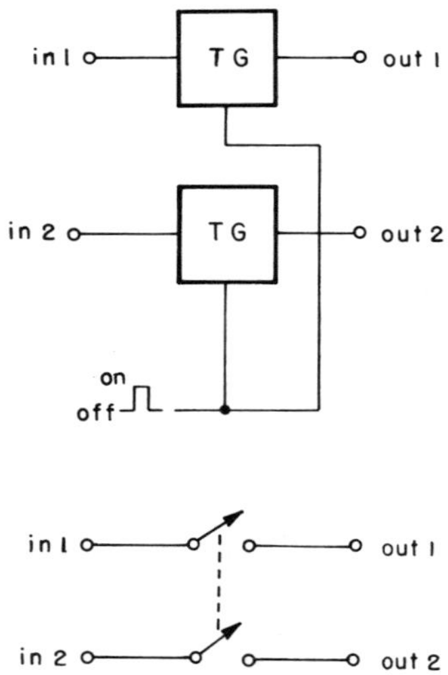

(C) Dpst switch.

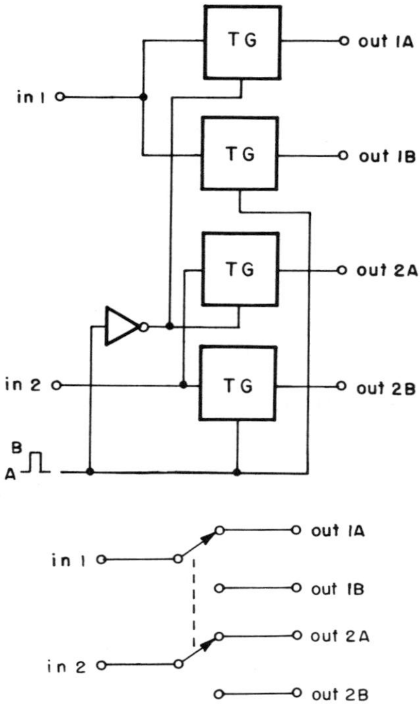

(D) Dpdt switch.

Fig. 4-3. Switch functions using transmission gates.

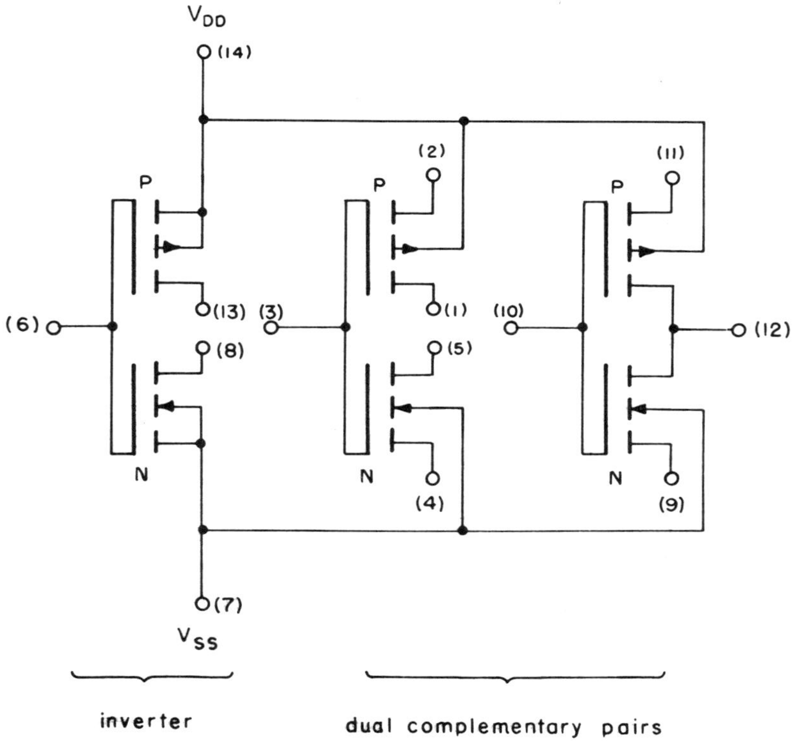

Fig. 4-4. 4007 dual complementary pair plus inverter schematic.

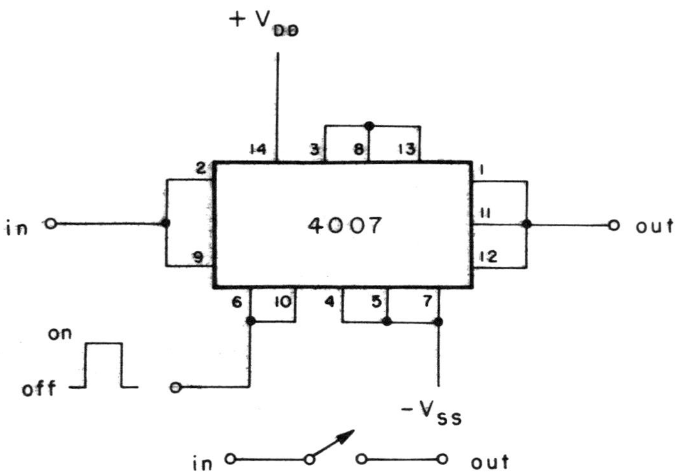

Fig. 4-5. Spst switch using a 4007.

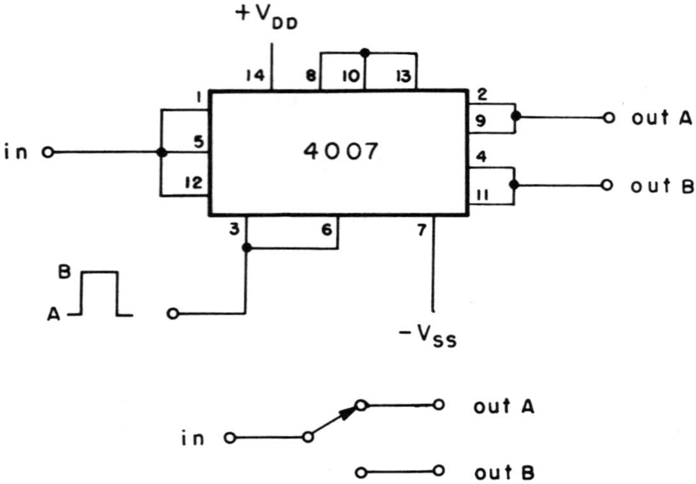

Fig. 4-6. Spdt switch using a 4007.

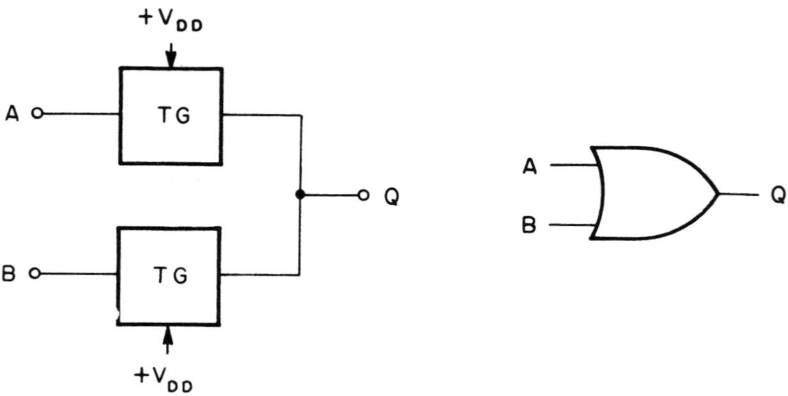

Fig. 4-7. A 2-input OR gate made with transmission gates.

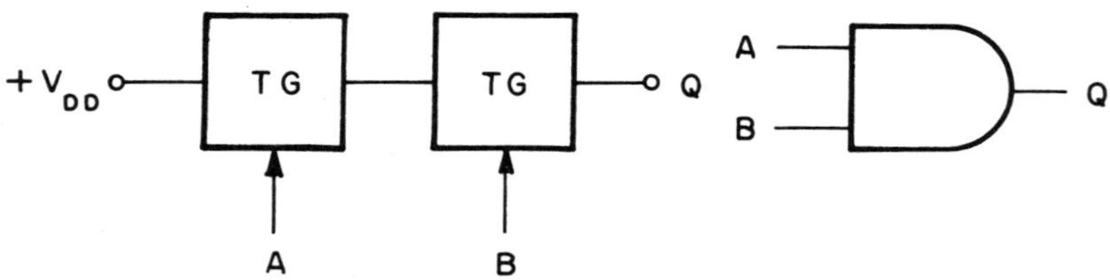

Fig. 4-8. A 2-input AND gate made with transmission gates.

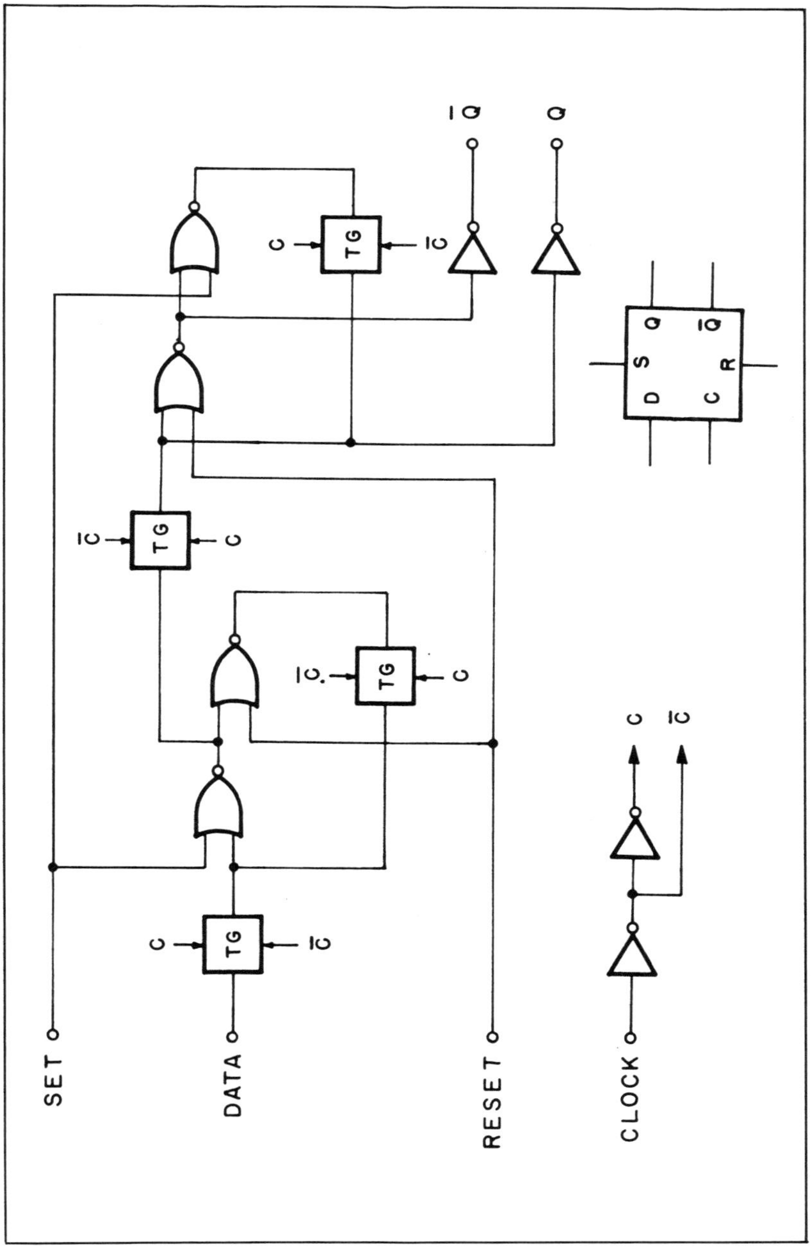

Fig. 4-9. D-type flip-flop.

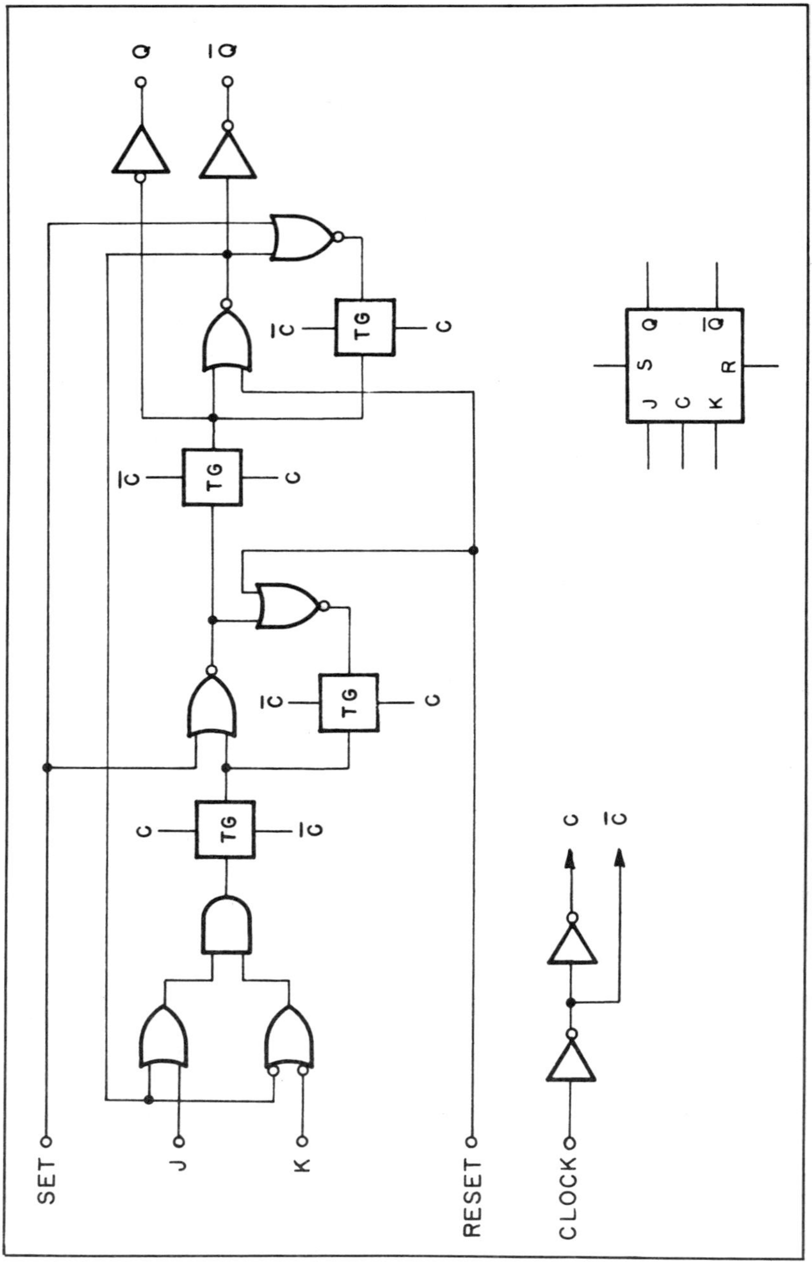

Fig. 4-10. JK flip-flop.

AND gates are also possible, as shown in Fig. 4-8. The output will be a logic 1 only if all inputs are also at logic 1, otherwise the output is at logic 0.

When used in conjunction with 4 NOR gates and 4 inverters, a master-slave D-type flip-flop can be formed, as shown in Fig. 4-9, and is identical to one-half of a 4013 device. Fig. 4-10 illustrates the use of transmission gates in the formation of a JK flip-flop (identical to one-half of a 4027).

Transmission Gates and Operational Amplifiers

Because transmission gates are able to accept analog as well as digital inputs, they lend themselves well to a number of circuits using operational amplifiers. As shown in Fig. 4-11, a series of transmission gates are used to control the gain of an inverting amplifier.

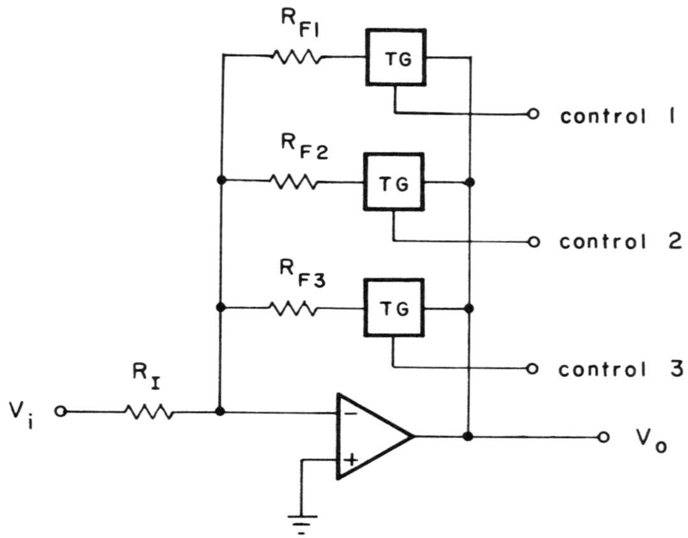

Fig. 4-11. Digital control of op-amp inverting amplifier closed-loop gain.

If the feedback resistance is made large enough, the on resistance of the transmission gate will have negligible effect on the gain, so that

$$\text{Gain} = \frac{V_o}{V_i} = \frac{R_F}{R_I} \qquad \text{(Eq. 4-1)}$$

It should be pointed out that both the op-amp and the transmission gate, such as a 4016, are powered with split supplies (±7.5 volts maximum). If the gain of this type of circuit is to be controlled by digital logic, such as a decoder operating between ground and V_{DD} (+7.5 volts maximum), the additional circuit of Fig. 4-12 is required for logic-level translation.

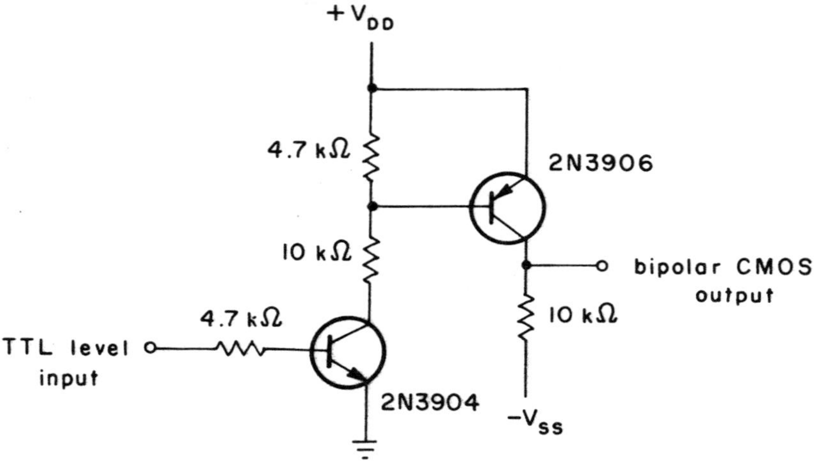

Fig. 4-12. Circuit for TTL to bipolar CMOS voltage level translation.

Fig. 4-13 illustrates how transmission gates are used to invert an analog signal. Gate C is used as a simple inverter, while the CONTROL pin of the unused gate is tied to V_{SS} to prevent possible damage to the entire package resulting from high power dissipation.

For certain types of active filters, transmission gates can be used to shift the center or cutoff frequency instead of using multielement potentiometers and voltage-variable resistors (i.e., transconductance amplifiers). Fig. 4-14 shows how two transmission gates are connected to a second order equal-component low-pass filter. The transmission gate is placed in series with each frequency determining resistor R. If the two switches are closed all of the time, the cutoff frequency of the filter is given by

$$f = \frac{1}{2\pi(RC)} \qquad \text{(Eq. 4-2)}$$

Now, if the control voltage of the two gates is switched on and off between V_{DD} and V_{SS} at, say, 50:50 rate (50% duty cycle), the effective, or average, resistance is then *twice* that of when the switch was closed all the time. Consequently, the cutoff frequency is now half the former value. Then by varying the duty cycle of the con-

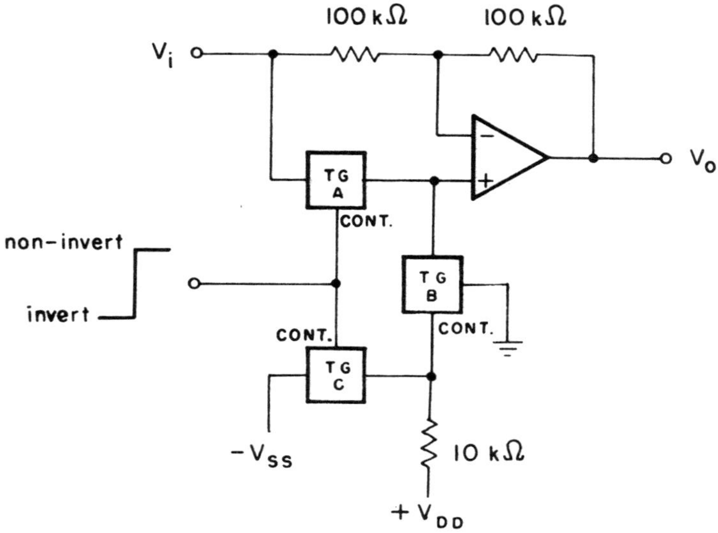

Fig. 4-13. Inverting an analog signal under digital control.

trol voltage to the pair of transmission gates, we can in turn shift the cutoff frequency of the filter. Mathematically,

$$f = \frac{1}{2\pi R_{eff} C} \qquad \text{(Eq. 4-3)}$$

where

$$R_{eff} = \frac{R}{\%\ \text{duty cycle}} \times 100$$

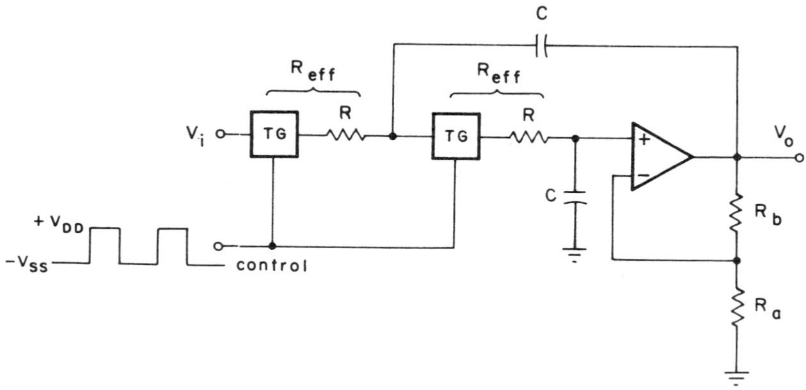

Fig. 4-14. Second-order low-pass active filter with variable cutoff frequency.

The frequency of the control voltage must be at least 10 times the highest cutoff frequency of the filter for proper operation. A constant frequency, variable duty cycle generator using a 555 timer is described in Chapter 5.

ANALOG MULTIPLEXERS/DEMULTIPLEXERS

CMOS combination multiplexers/demultiplexers are basically composed of a series of transmission gates with some kind of BCD decoder, as shown in the block diagram of Fig. 4-15. Instead of constructing our own, CMOS manufacturers produce a variety of multiplexers/demultiplexers, such as:

- 4051 single 8-channel multiplexer/demultiplexer
- 4052 differential 4-channel multiplexer/demultiplexer
- 4053 triple 2-channel multiplexer/demultiplexer
- 4067 single 16-channel multiplexer/demultiplexer
- 4097 differential 8-channel multiplexer/demultiplexer

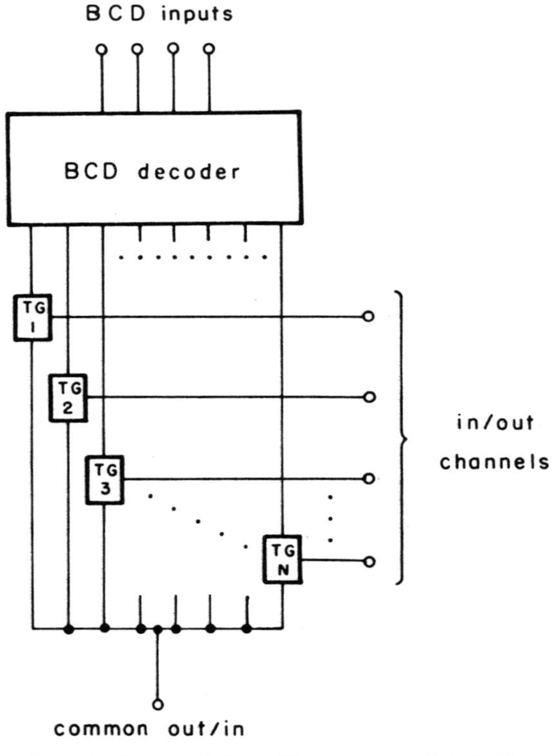

Fig. 4-15. Analog multiplexer/demultiplexer block diagram.

with the corresponding pin diagrams shown in Fig. 4-16. For the 4051, 4052, and 4053 devices, the on resistance is typically 150 Ω, while it is 200 Ω for the 4067 and 4097. For all the devices listed above, channel selection is inhibited when the INHIBIT input is at logic 1. When the INHIBIT input is at logic 0, channel selection is determined by a 2-bit binary word for the 4052, a 3-bit word for the 4051, 4053, 4097, and a 4-bit word for the 4067 (Table 4-1).

Table 4-1. Channel Select Code

Channel	D	C	B	A
0	0	0	0	0
1	0	0	0	1
2	0	0	1	0
3	0	0	1	1
4	0	1	0	0
5	0	1	0	1
6	0	1	1	0
7	0	1	1	1
8	1	0	0	0
9	1	0	0	1
10	1	0	1	0
11	1	0	1	1
12	1	1	0	0
13	1	1	0	1
14	1	1	1	1
15	1	1	1	1

The V_{EE} pins of the 4051, 4052, and 4053 are connected to the V_{SS} pins for digital signals, and to a negative supply voltage ($V_{DD} - V_{EE} = 15$ volts maximum) while V_{SS} is at ground for analog signals. For the 4067 or 4097, the V_{SS} pins are connected in a like manner for the V_{EE} pin of the 4051, 4052, or 4053.

AN INTRODUCTION TO THE EXPERIMENTS

The following experiments are designed to demonstrate the operation and application of CMOS transmission gates and multiplexers/demultiplexers. The experiments that you will perform can be summarized as follows:

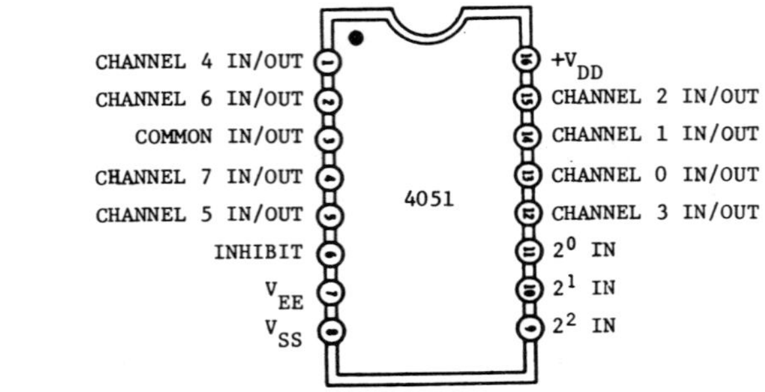

(A) 4051 single 8-channel multiplexer/demultiplexer.

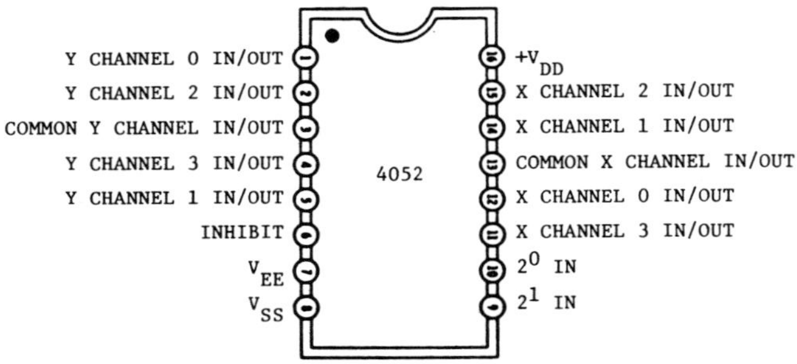

(B) 4052 differential 4-channel multiplexer/demultiplexer.

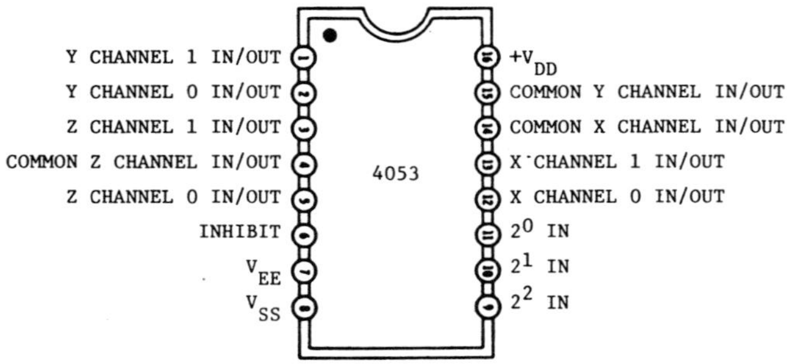

(C) 4053 triple 2-channel multiplexer/demultiplexer.

Fig. 4-16. CMOS analog

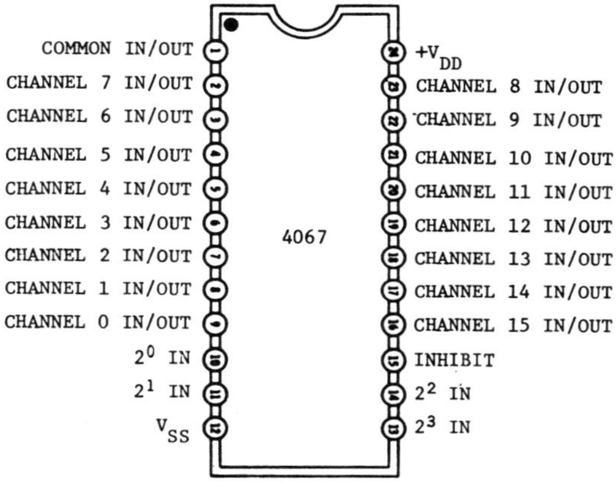

(D) 4067 single 16-channel multiplexer/demultiplexer.

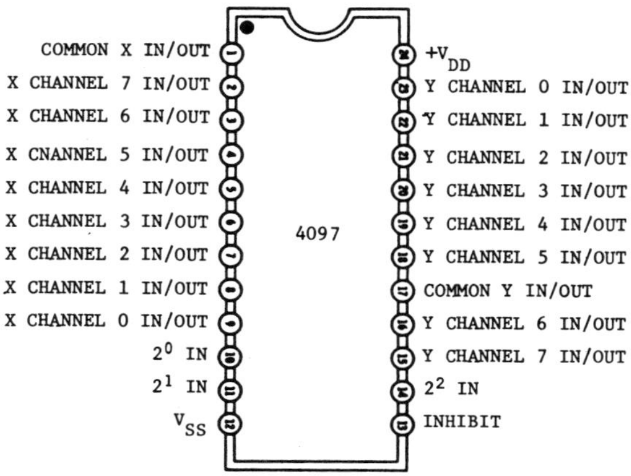

(E) 4097 differential 8-channel multiplexer/demultiplexer.

multiplexer/demultiplexer pin diagrams.

Experiment No.	*Purpose*
1	Demonstrates the operation of a single 4016 transmission gate, or bilateral switch.
2	Demonstrates how transmission gates are connected to represent an spdt switch.
3	Demonstrates the operation of a 4051 single 8-channel multiplexer/demultiplexer.

EXPERIMENT 1

Purpose

The purpose of this experiment is to demonstrate the operation of a single transmission gate, or bilateral switch, using the 4016.

Pin Configuration of Integrated Circuit Chip (Fig. 4-17)

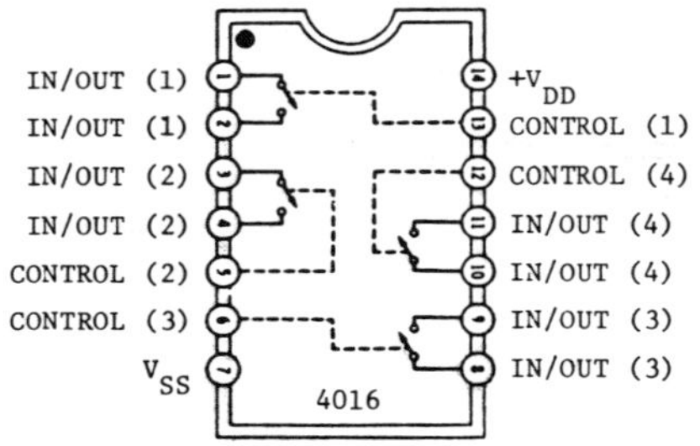

Fig. 4-17.

Schematic Diagram of Circuit (Fig. 4-18)

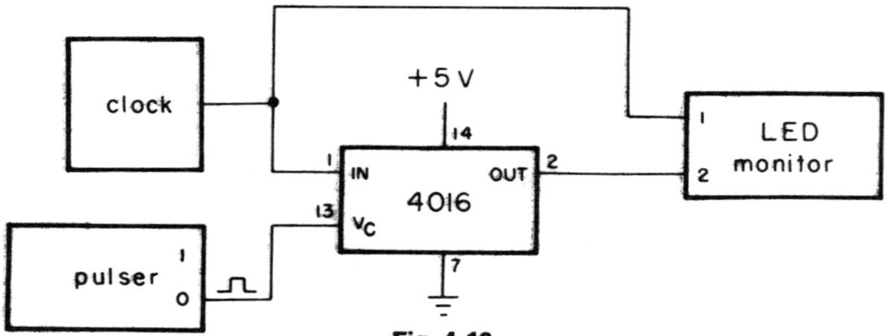

Fig. 4-18.

Step 1

Wire the circuit as shown in the schematic diagram. Since all unused CMOS inputs cannot be left floating, connect pins 4, 5, 6, 8, 10, and 11 to ground.

Step 2

Apply power to the breadboard and adjust the output frequency of the clock so that LED monitor No. 1 flashes approximately once each second. What do you observe about LED monitor No. 2?

Since the CONTROL input is at logic 0, the transmission gate acts as an open circuit from input to output (pin 1 to 2). Consequently, the LED monitor should be *unlit.*

Step 3

Now press in the pulser so that the CONTROL input of the transmission gate is at logic 1. What do you now observe happening to LED monitor No. 2?

You should observe that LED monitor No. 2 is flashing exactly in the same manner as LED monitor No. 1, which is the input signal. Since the CONTROL input is at logic 1, the transmission gate behaves as a short circuit from input to output.

Release the pulser. You should observe that LED monitor No. 2 is now unlit, as in Step 2, so that the transmission gate behaves as an open circuit when the CONTROL input is at logic 0.

Step 4

Disconnect the power from the breadboard. Now reverse the input and output connections to the transmission gate so that the clock and LED monitor No. 1 are now connected to pin 2, while LED monitor No. 2 is connected to pin 1 of the 4016. Repeat Steps 2 and 3. Do you notice any differences?

There should be no difference! Consequently, it makes no difference which is the input or output of a transmission gate since they are interchangeable. Another name for the transmission gate is a *bilateral switch,* which can be used to switch analog as well as digital signals. When the CONTROL input is at logic 1, the switch does not represent a true short circuit, but rather a finite "on" resistance. For the 4016 transmission gate, this resistance is typically 300 ohms from input to output.

EXPERIMENT 2

Purpose

The purpose of this experiment is to demonstrate how transmission gates can be connected to simulate a single-pole-double-throw (spdt) switch.

Pin Configuration of Integrated Circuit Chips (Fig. 4-19)

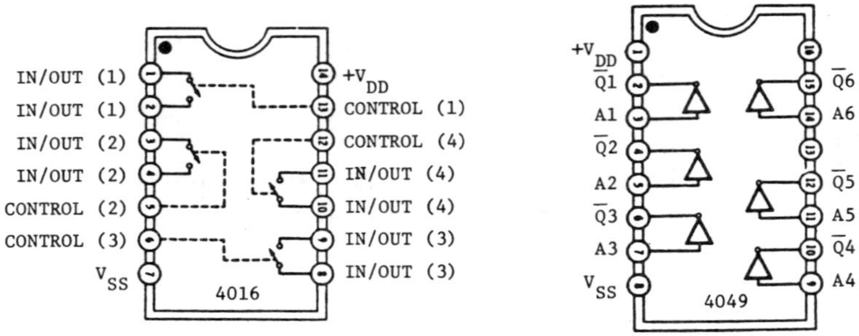

Fig. 4-19.

Schematic Diagram of Circuit (Fig. 4-20)

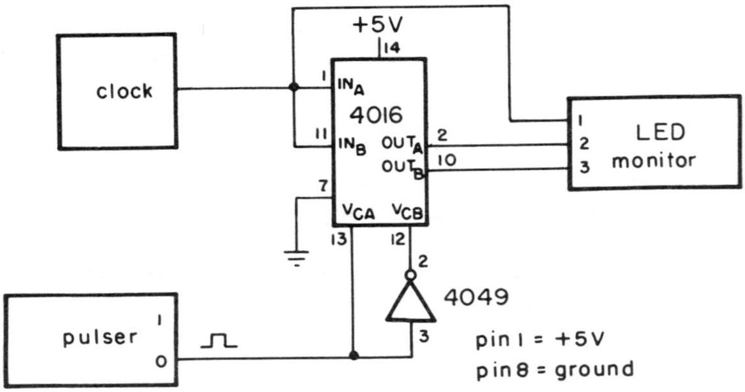

Fig. 4-20.

Step 1

Wire the circuit as shown in the schematic diagram. Although not specifically shown in the schematic, make sure that the V_{DD} pins of both the 4016 (pin 14) and 4049 (pin 1) are connected to the positive supply voltage and the V_{SS} pins (pin 7 of 4016 and pin

8 of 4049) are connected to ground. Since all unused inputs have to be terminated, connect pins 5, 7, 9, 11, and 14 of the 4049 hex inverter to ground and pins 3, 5, 6, and 8 of the 4016 to ground.

Step 2

Apply power to the breadboard, and adjust the output frequency of the clock so that LED monitor No. 1 flashes approximately once each second. What is happening to LED monitors No. 2 and No. 3?

You should observe that LED monitor No. 2 is flashing exactly like LED monitor No. 1, while LED monitor No. 3 is unlit.

Step 3

Press in the pulser. What difference do you notice now concerning the LED monitors?

You should have observed that LED monitor No. 3 is now flashing exactly in step with LED monitor No. 1, while LED monitor No. 2 is now unlit. Release the pulser. What changes do you observe?

As in Step 2, LED monitor No. 2 is flashing in step with LED monitor No. 1, while LED monitor No. 3 is unlit. Depending on the state of the output of the pulser, we are able to switch the input signal to either one of two output terminals, which is precisely the operation of an spdt *mechanical* switch.

EXPERIMENT 3

Purpose

The purpose of this experiment is to demonstrate the operation of the 4051 multiplexer/demultiplexer.

Step 1

Wire the circuit as shown in the schematic diagram and apply power to the breadboard. Set the three logic switches all at logic 0. Which LED monitor would you predict to be flashing on and off, corresponding to the clock input?

Pin Configuration of Integrated Circuit Chip (Fig. 4-21)

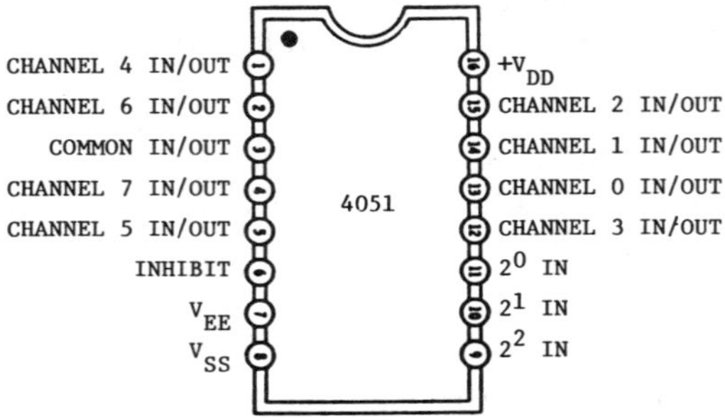

Fig. 4-21.

Schematic Diagram of Circuit (Fig. 4-22)

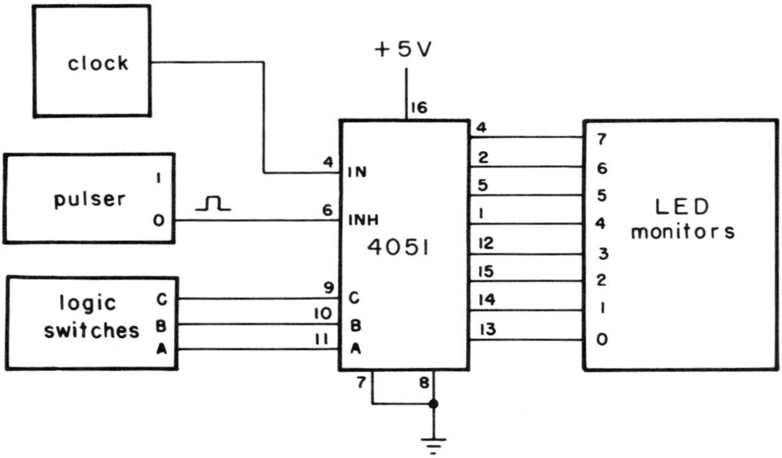

Fig. 4-22.

LED monitor 0 should be flashing since the 3-bit code is set at CBA = 000 which corresponds to the decimal number 0.

Step 2

Change the settings of the logic switches according to Table 4-2, filling in your results.

You should have observed that the flashing LED monitor corresponds to the decimal equivalent of the 3-bit code set by the logic switches. Consequently, the 4051 allows us to transmit input data to any one of eight output channels, commonly called an *8-channel demultiplexer*.

Table 4-2. Logic Switch Setting Determines Flashing Output Channel

Logic Switch Setting C B A	Flashing Output Channel
0 0 0	0
1 1 0	
0 1 0	
1 0 1	
0 1 1	
1 1 1	
1 0 0	
0 0 1	

Step 3

At this point, LED monitor No. 1 should be flashing. Now hold in the pulser that is connected to pin 6 of the 4051. What happens?

LED monitor No. 1 ceases to flash. Now release the pulser. What happens?

The LED monitor should flash on and off as before. When the INHIBIT pin is at logic 1, the input data is prevented from being transmitted to the output channel selected by the logic switches.

Step 4

A *multiplexer* performs the opposite function of a demultiplexer, in that we are able to select from one of several inputs and transmit that input data to a single output channel. Using the 4051, determine how you would connect it to function as an 8-channel multiplexer. In addition describe how you might automatically sequence all eight input channels.

CHAPTER 5

Monostable and Astable Multivibrators

INTRODUCTION AND OBJECTIVES

Multivibrators primarily provide the "heartbeat" for most digital circuits. In this chapter a number of useful circuits that permit one-shot free-running, as well as voltage-controlled frequency operation, are discussed. Included will be circuits using standard logic elements as well as specialized monolithic devices.

At the completion of this chapter, you will be able to do the following:

- Construct monostable and astable circuits using standard logic elements, such as inverters, NOR gates, and NAND gates.
- Construct monostable circuits using the following devices:
  - 4047 monostable/astable multivibrator
  - 4098 dual monostable multivibrator
  - 74C221 dual monostable multivibrator
  - 555 timer
- Construct astable circuits using the following devices:
  - 4007 dual complementary pair plus inverter
  - 4047 monostable/astable multivibrator
  - 555 timer
- Construct astable circuits with variable duty cycle using the following devices:
  - 4049 hex inverter
  - 555 timer

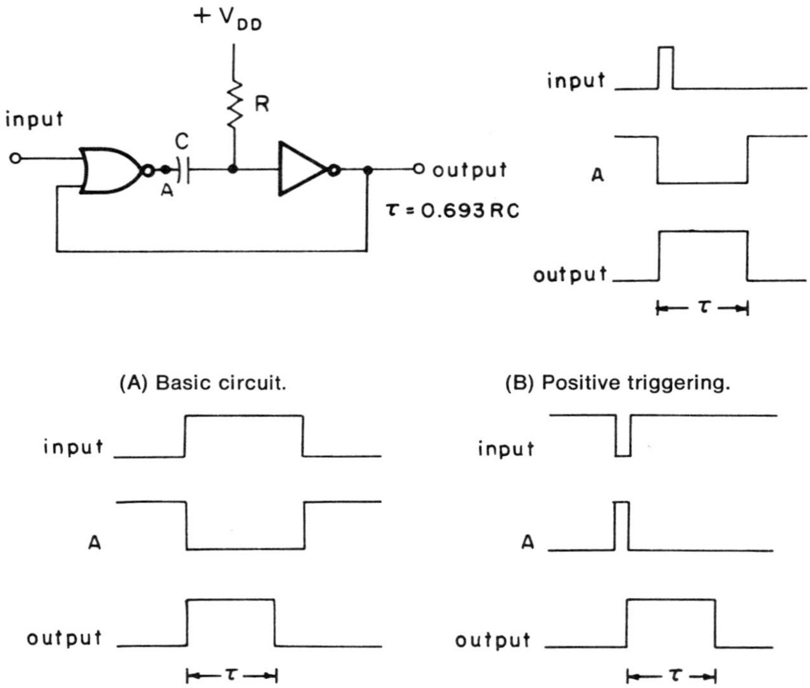

(A) Basic circuit.

(B) Positive triggering.

(C) Positive trigger pulse width longer than output pulse width.

(D) Negative triggering.

Fig. 5-1. Monostable multivibrator using a 2-input NOR gate.

DISCRETE MONOSTABLE MULTIVIBRATORS

In this section several circuits are briefly discussed using standard logic gates and having time delays of acceptable accuracy. As shown in Fig. 5-1, a 2-input NOR gate, such as a 4001 or 74C02, with an inverter can be wired in a simple circuit having a single time constant.

For a positive trigger pulse, the output pulse from the inverter will also be positive for a specified time period τ (in seconds) such that

$$\tau = -RC\left[ln\left(1 - \frac{V_{TR}}{V_{DD}}\right)\right] \qquad \text{(Eq. 5-1)}$$

where V_{TR} is the transfer voltage threshold at which the inverter switches from one logic level to the other. Ideally V_{TR} is one-half the supply voltage V_{DD}, so that Equation 5-1 simplifies to

$$\tau = 0.693(RC) \text{ second} \qquad \text{(Eq. 5-2)}$$

However, the transfer threshold may vary from 30% to 70% of V_{DD} (±40% change) with different units. This possible variation, in turn, varies the output pulse width anywhere from −48% to +74% of the ideal value. As illustrated in the timing diagram of Fig. 5-1B, the circuit responds to the *positive-going leading edge* of the input trigger pulse. If the trigger pulse width is shorter than the output pulse, the output of the NOR gate is the complement of the output of the inverter. If the input pulse width is *longer* than the output (Fig. 5-1C), the output of the inverter is a positive pulse for a duration given by Equation 5-1 or 5-2, but the output of the NOR gate is now the complement of the input trigger.

Fig. 5-1D illustrates the timing diagram for a negative trigger pulse. Note that the monostable timing period does not start until the *positive-going trailing edge* of the trigger occurs. For this NOR gate monostable, the output of the NOR gate is the complement of the negative trigger pulse.

A 2-input NAND gate, such as a 4011 or 74C00, can be used to form a monostable circuit, as shown in Fig. 5-2, which is primarily

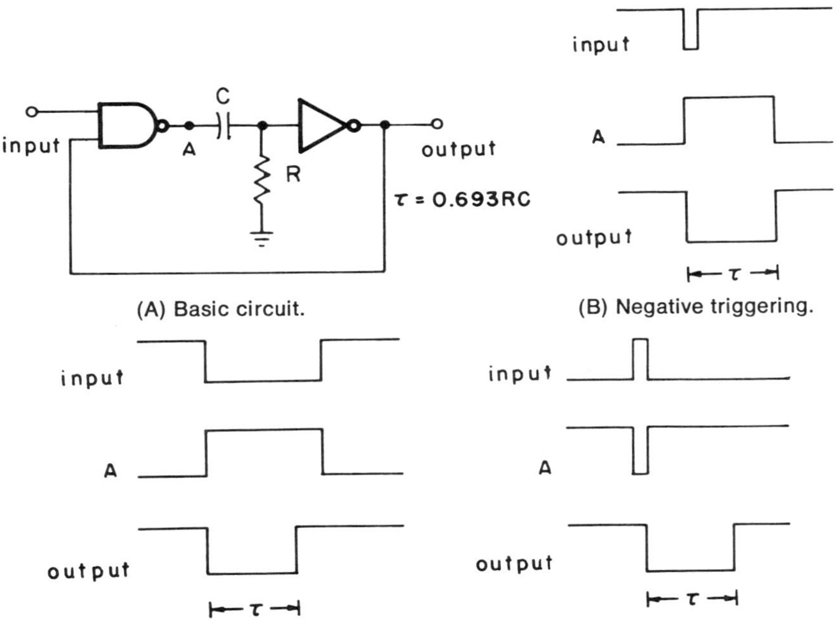

(A) Basic circuit.

(B) Negative triggering.

(C) Negative trigger pulse width longer than output pulse width.

(D) Positive triggering.

Fig. 5-2. Monostable multivibrator using a 2-input NAND gate.

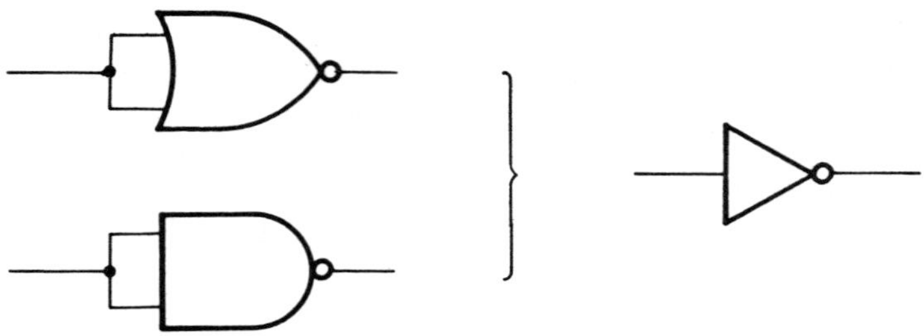

Fig. 5-3. NOR and NAND gates as inverters.

used for negative triggering. The inverter used in Figs. 5-1 and 5-2 is equivalent to either a 2-input NOR gate, or a 2-input NAND gate with both inputs tied together, as shown in Fig. 5-3. Consequently, the inverter shown in Fig. 5-1 is conveniently replaced with another NOR gate, while the circuit of Fig. 5-2 would use a second NAND gate in place of the inverter.

4047 MONOSTABLE

Instead of constructing discrete monostable circuits where the output pulse width might be unpredictable, it might be more advantageous to consider one of several monolithic devices as a suitable alternative. One such device is the 4047, whose pin configuration is shown in Fig. 5-4. The 4047 monostable possesses a higher degree of accuracy and stability when compared with the discrete monostable circuits, in addition to lower power dissipation.

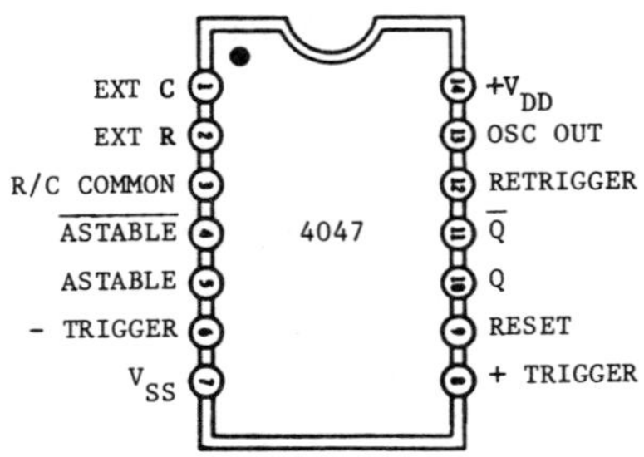

Fig. 5-4. Pin diagram of the 4047 multivibrator.

Positive-edge triggering is accomplished using the circuit of Fig. 5-5, while the negative-edge trigger circuit is shown in Fig. 5-6. In either case, the minimum trigger pulse width is typically 500 ns for a 5-volt supply; 200 ns for a 10-volt supply; and 140 ns for 15 volts.

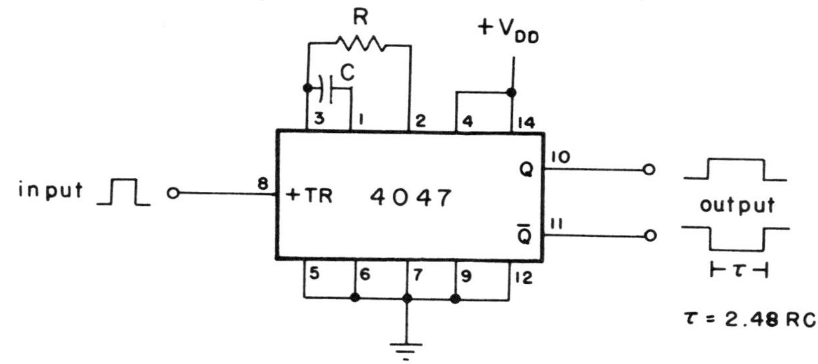

Fig. 5-5. 4047 monostable wired for positive triggering.

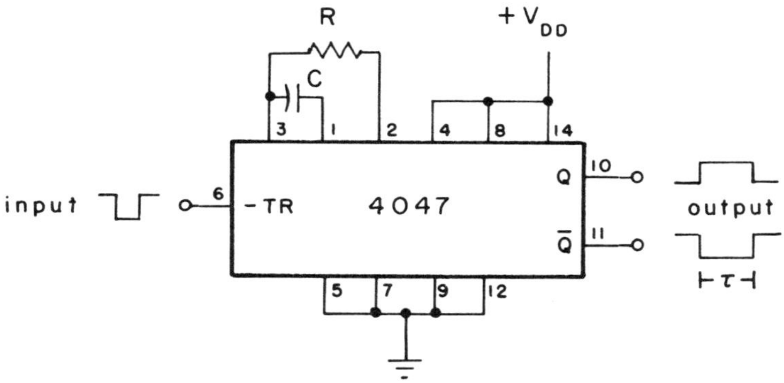

Fig. 5-6. 4047 monostable wired for negative triggering.

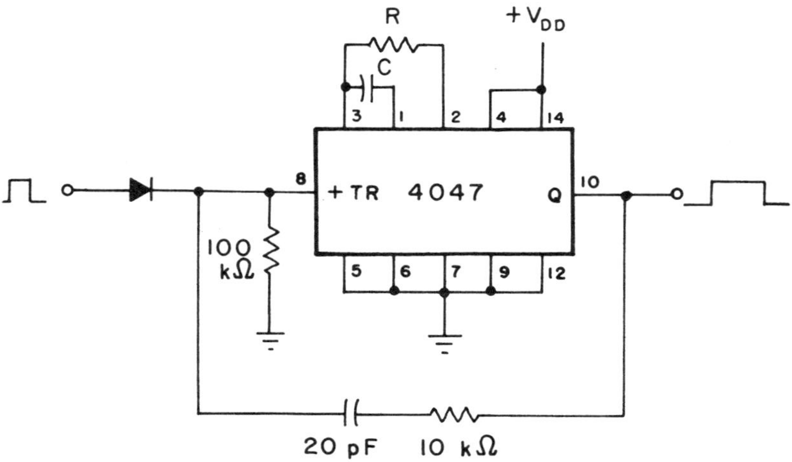

Fig. 5-7. Input pulse stretcher circuit.

When triggered, the pulse width for the complementary outputs Q and $\overline{Q}$ is given by

$$\tau = 2.48(RC) \text{ seconds} \qquad \text{(Eq. 5-3)}$$

However, for reliable operation, the timing resistor should be in the 10 kΩ to 1 MΩ range, while the timing capacitor can be any practical value greater than 0.001 μF. Consequently, the output pulse width can be as small as 25 μs.

In some cases input pulses shorter than the minimum required for triggering (e.g., noise transients) may cause erratic operation. The circuit shown in Fig. 5-7 circumvents this possibility by basically "stretching" the input pulse via the 20 pF-10 kΩ feedback path.

555 TIMER MONOSTABLE

Since its introduction, the 555 integrated circuit timer has been one of the most popular and versatile integrated circuit devices in use. Although compatible with CMOS circuits, the 555 timer, nevertheless, may be unsuitable for battery-powered equipment because of its current drain. However, there have recently appeared several low-power pin-for-pin equivalents.

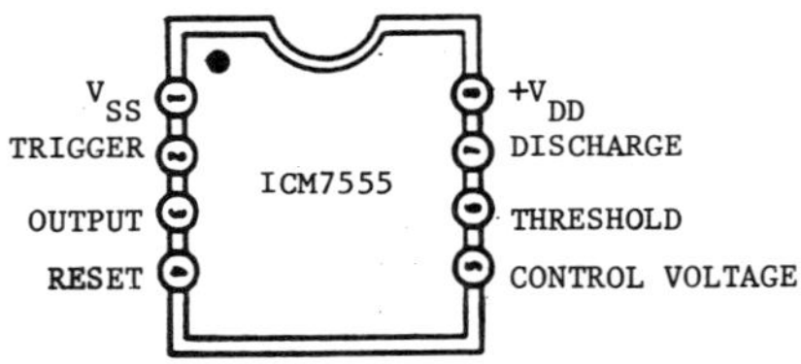

Fig. 5-8. Pin diagram of the ICM7555 timer.

One such equivalent is the ICM7555 CMOS timer (Intersil), whose pin configuration is illustrated in Fig. 5-8. This CMOS device has a typical power supply current rating of 80 μA, less than 5% of that required by the standard bipolar 555 timer. In addition, the power supply voltage can range from 2 to 18 volts.

When wired for monostable operation, as shown in Fig. 5-9, the timer is triggered by a negative pulse. Based upon the selection of the external timing resistor and capacitor, the output at pin 3 will be positive for a time period τ (in seconds) so that

$$\tau = 1.1(RC) \text{ seconds} \qquad \text{(Eq. 5-4)}$$

For reliable operation, the timing resistor should be in the 1-kΩ to 10-MΩ range, while the timing capacitor can be any practical value greater than 0.001 μF.

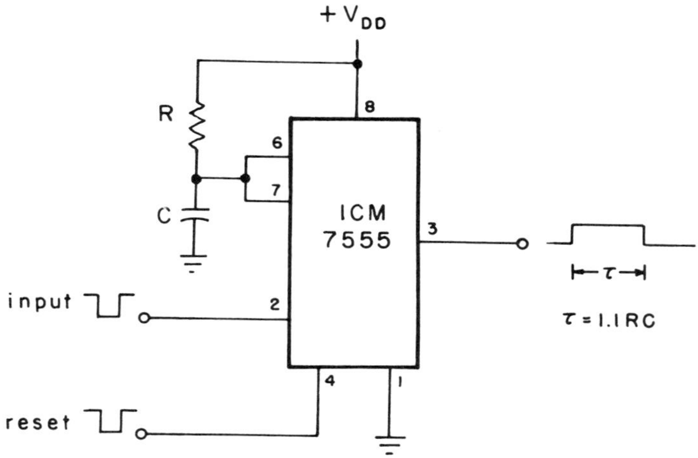

Fig. 5-9. ICM7555 monostable circuit.

Once triggered the output will remain at logic 1 until the set time has elapsed, even if it is triggered again during this interval. If, on the other hand, a negative-going pulse is applied to the reset terminal (pin 4) during the timing cycle, the timing cycle is halted and the output returns to logic 0. The output of the ICM7555 can supply up to 100 mA, and, thus, is capable of sourcing or sinking TTL as well as CMOS loads. The XR-L555 timer (EXAR) is another low-power version that is currently available. More detailed notes and applications using the 555 timer can be found in *The 555 Timer Applications Sourcebook, with Experiments* published by Howard W. Sams & Co., Inc.

OTHER MONOLITHIC DEVICES

Besides the 555 timer and the 4047, there are two other monolithic devices primarily intended for monostable operation. One is the 4098 (RCA), or the equivalent MC14528 (Motorola), and the other is the 74C221 (National Semiconductor). Both devices are *dual* monostable multivibrators contained in a single package, and are identical in function to the 74123 TTL unit.

As shown in the pin diagram of Fig. 5-10, the 4098 (or MC14528) has both a leading-edge (+TR) and a trailing-edge (−TR) input in addition to complementary outputs Q and $\overline{Q}$. Fig. 5-11 shows the connections required for positive-edge triggering of both sections. Once triggered, the output pulse width is given by

$$\tau = 0.5(RC) \text{ second} \qquad \text{(Eq. 5-5)}$$

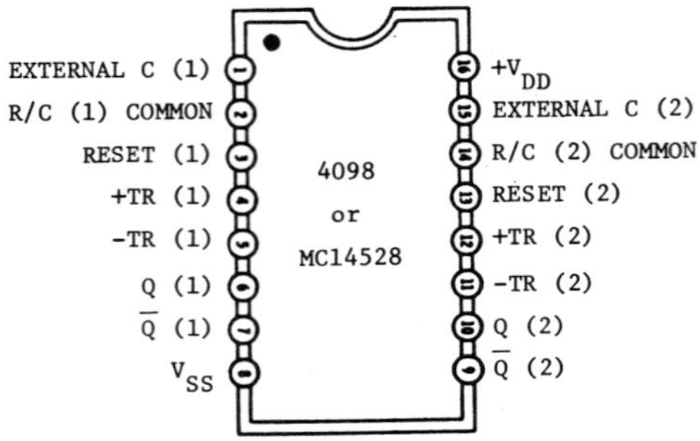

Fig. 5-10. Pin diagram of the 4098/MC14528 dual monostable multivibrator.

for the external timing capacitor in the 0.01 to 100 μF range. The minimum value for the external timing resistor is recommended at 5 kΩ. Even if the circuit is retriggered repeatedly during the timing cycle, the timing cycle will go to completion as if triggered only by a single pulse. However, if a negative-going pulse is applied to the RESET pin, the timing cycle is immediately terminated. If the reset function is not used, it should be connected to V_{DD}. If, on the other hand, an entire section is not used for a particular application, the RESET and +TR pins are tied to V_{SS}, while the −TR input is connected to V_{DD}. Fig. 5-12 shows the required connections for triggering the 4098 with a series of positive leading-edge pulses. In this mode, Equation 5-5 determines the output pulse width *after* the last triggering edge is encountered.

The connections required for trailing-edge triggering are shown in Figs. 5-13 and 5-14. Taking into account the extreme values allowed for the external timing components, Table 5-1 summarizes

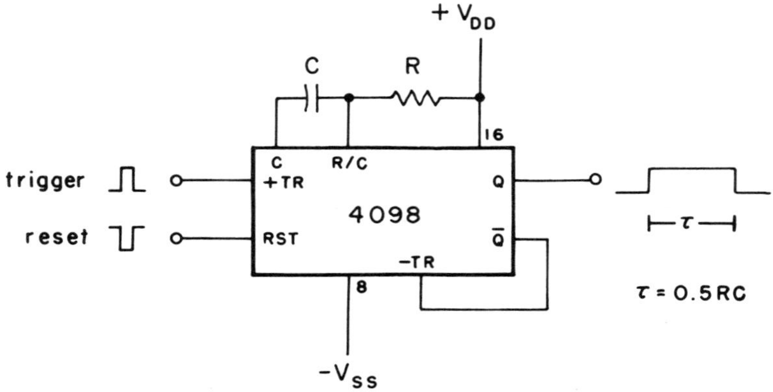

Fig. 5-11. 4098 positive nonretriggerable monostable.

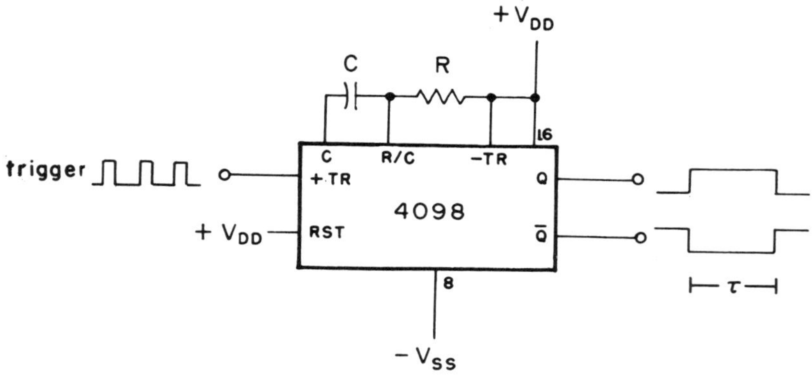

Fig. 5-12. 4098 positive retriggerable monostable.

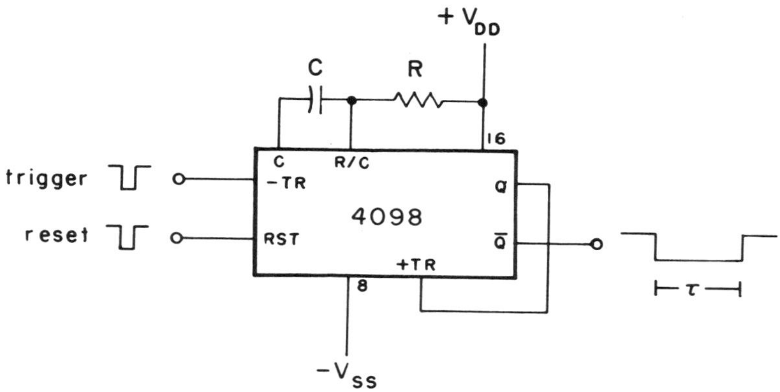

Fig. 5-13. 4098 negative nonretriggerable monostable.

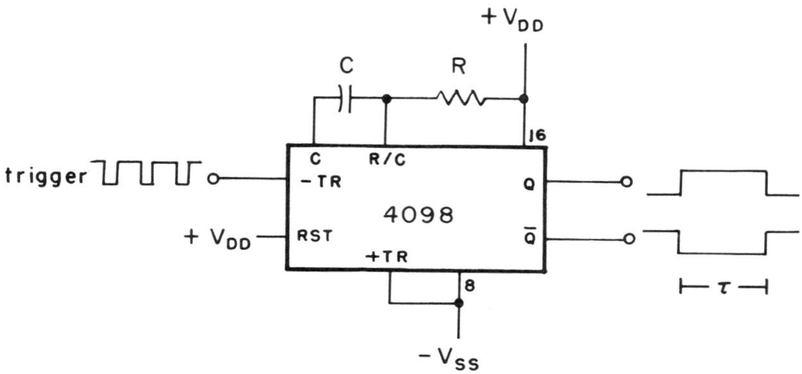

Fig. 5-14. 4098 negative retriggerable monostable.

Table 5-1. Typical Minimum Allowable Trigger and Reset Pulse Widths

Supply Voltage	Trigger	Reset
5 V	70 ns	25 μs
10 V	30 ns	15 μs
15 V	20 ns	10 μs

the typical minimum allowable input trigger and reset pulse widths for the 4098.

The 74C221, unlike the 4098, is an exact pin-for-pin CMOS equivalent of the 74123 TTL dual monostable, as shown by the pin dia-

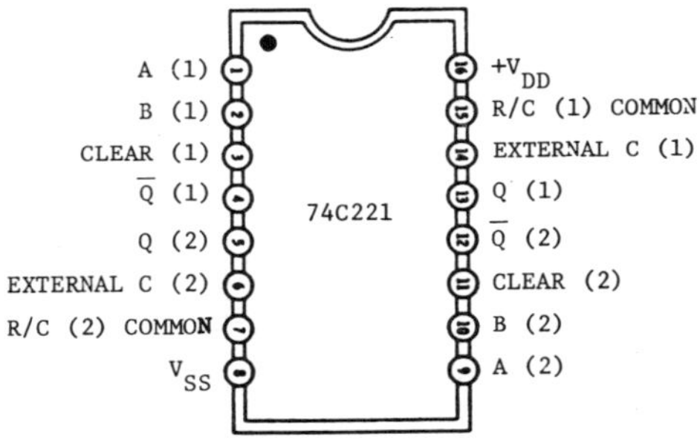

Fig. 5-15. Pin diagram of the 74C221 dual retriggerable monostable.

gram of Fig. 5-15. However, the pulse width equations are not the same. For the 74C221, the output pulse width is given by

$$\tau = 0.994(\mathrm{RC}) \text{ second} \qquad \text{(Eq. 5-6)}$$

which is frequently approximated by $\tau = \mathrm{RC}$ for all practical purposes.

DISCRETE ASTABLE MULTIVIBRATORS

As shown in Fig. 5-16, one of the simplest types of CMOS astable multivibrators uses just two inverters, such as a 4049 or 74C04, with a single time constant. For this circuit, the output frequency depends on the transfer voltage threshold V_{TR}, so that

$$f = \frac{1}{T}$$

$$= \frac{1}{t_1 + t_2} \qquad \text{(Eq. 5-7A)}$$

where

$$t_1 = -RC\ ln\left(1 - \frac{V_{TR}}{V_{DD}}\right) \qquad \text{(Eq. 5-7B)}$$

and

$$t_2 = -RC\ ln\left(\frac{V_{TR}}{V_{DD}}\right) \qquad \text{(Eq. 5-7C)}$$

The time period during which the output is high is t_1, while t_2 is the time for which the output is low. By combining Equations 5-7B and 5-7C, the period of oscillation is then written as

$$T = -RC\ ln\left[\left(1 - \frac{V_{TR}}{V_{DD}}\right)\left(\frac{V_{TR}}{V_{DD}}\right)\right] \qquad \text{(Eq. 5-8)}$$

Since V_{TR} is ideally one-half the supply voltage, t_1 equals t_2, or a 50% duty cycle. In addition, Equation 5-7A when replaced with Equation 5-8 ideally reduces to

$$f = \frac{0.721}{RC}\ \text{Hz} \qquad \text{(Eq. 5-9)}$$

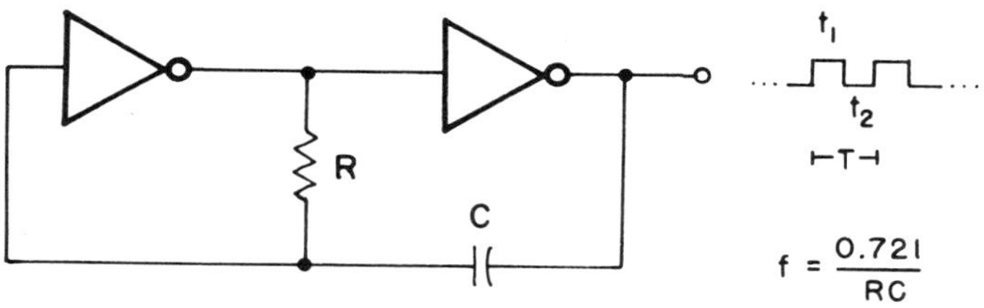

Fig. 5-16. CMOS astable multivibrator.

Because CMOS technology now typically uses three diodes (Fig. 5-17) to protect the input MOS gates from damage, the output frequency of the circuit of Fig. 5-16 is sensitive to supply voltage changes. One effective method to minimize this sensitivity is to add a second resistor R_S as shown in Fig. 5-18. Normally, R_S is from 2 to 10 times the value of R, so that it will have no effect on the output frequency, as given by

$$f = \frac{1}{T} \qquad \text{(Eq. 5-10A)}$$

where

$$T = -RC\ ln\left[\left(\frac{V_{TR}}{V_{DD} + V_{TR}}\right)\left(\frac{V_{DD} - V_{TR}}{2V_{DD} - V_{TR}}\right)\right] \qquad \text{(Eq. 5-10B)}$$

Again assuming that $V_{TR} = 0.5V_{DD}$, Equation 5-10A simplifies to

$$f = \frac{0.455}{RC} \qquad \text{(Eq. 5-11)}$$

The timing resistor R is restricted to the 5-kΩ to 1-MΩ range, while C can be any value greater than 100 pF for reliable operation. Fig. 5-19 illustrates how the circuit of Fig. 5-18 can be used with NOR or NAND gates.

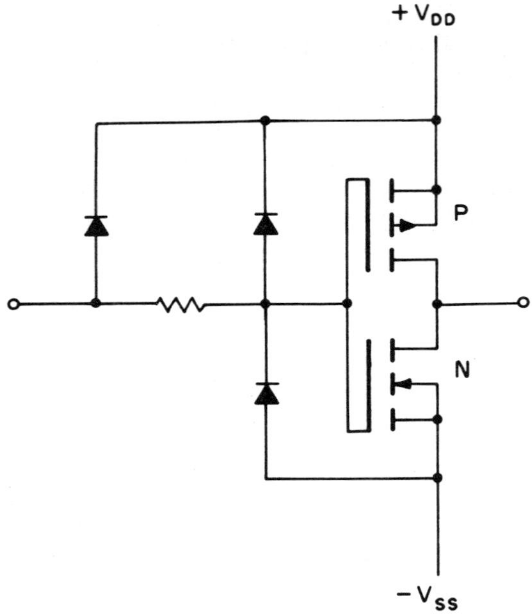

Fig. 5-17. CMOS internal input protection network.

A simpler astable circuit is the single Schmitt trigger astable with a single time constant, shown in Fig. 5-20. The output frequency (in hertz) is determined from the basic relationship

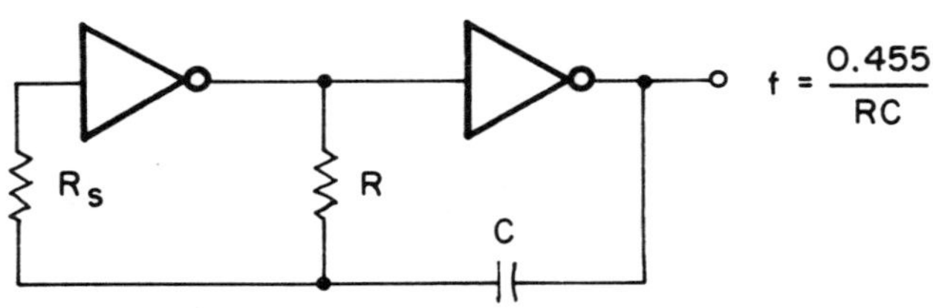

Fig. 5-18. Improved astable multivibrator.

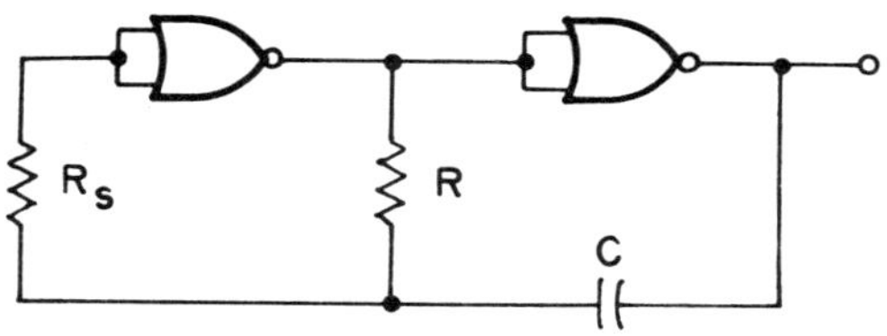

(A) Using NOR gates.

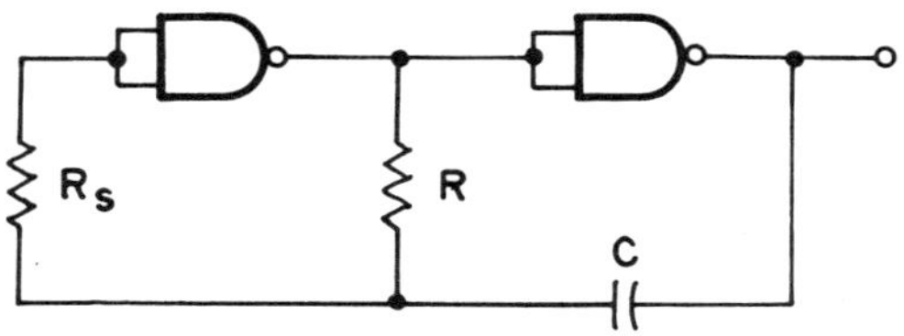

(B) Using NAND gates.

Fig. 5-19. Other simple astable circuits.

$$f = \frac{1}{RC\ ln\left[\frac{V_2(V_{DD} - V_1)}{V_1(V_{DD} - V_2)}\right]} \qquad \text{(Eq. 5-12)}$$

where V_1 and V_2 are the lower and upper threshold voltages of the Schmitt trigger, respectively. However, these thresholds are sensitive to changes in the supply voltage. Table 5-2 illustrates this sensitivity with the comparison of several Schmitt triggers from different

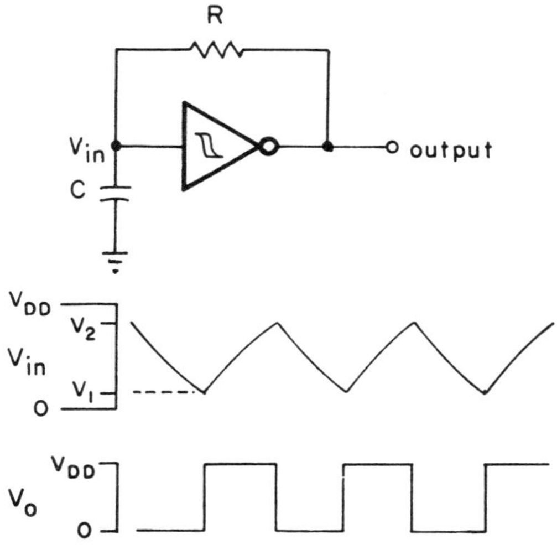

Fig. 5-20. Schmitt trigger astable.

Table 5-2. CMOS Schmitt Trigger Threshold Voltages

Device & Manufacturer	V_{DD}	V_1	V_2
74C14 (National Semiconductor)	5	1.4	3.6
	10	3.2	6.8
40106 (RCA)	5	1.9	2.9
	10	3.9	5.9
40106 (National Semiconductor)	5	1.4	3.6
	10	3.2	6.8
MC14584 (Motorola)	5	2.3	2.9
	10	3.9	5.9

These values are typical values from the above manufacturers' data sheets.

manufacturers. Even for devices having the same number, such as the 40106 Schmitt trigger, the voltage thresholds are different with different manufacturers. The threshold sensitivity can be clearly seen by rewriting Equation 5-12 in the general form:

$$f = \frac{\kappa}{RC} \qquad \text{(Eq. 5-13)}$$

where the constant κ is computed from the values of Table 5-2 and summarized in Table 5-3.

Table 5-3. Values for κ in Equation 5-13

Device & Manufacturer	Supply Voltage	
	5 V	10 V
74C14 (National Semiconductor)	0.529	0.663
40106 (RCA)	1.231	1.233
40106 (National Semiconductor)	0.529	0.663
MC14584 (Motorola)	2.070	1.233

4047 ASTABLE

As illustrated in Fig. 5-21, the 4047 is connected as an astable multivibrator having two output frequencies. When the output is taken from the OSC OUTPUT (pin 13), the nominal output frequency (in hertz) is determined from the equation

$$f = \frac{1}{2.2RC} \qquad \text{(Eq. 5-14)}$$

However, the output waveform at pin 13 will not necessarily have a 50% duty cycle. On the other hand, a symmetrical square wave can

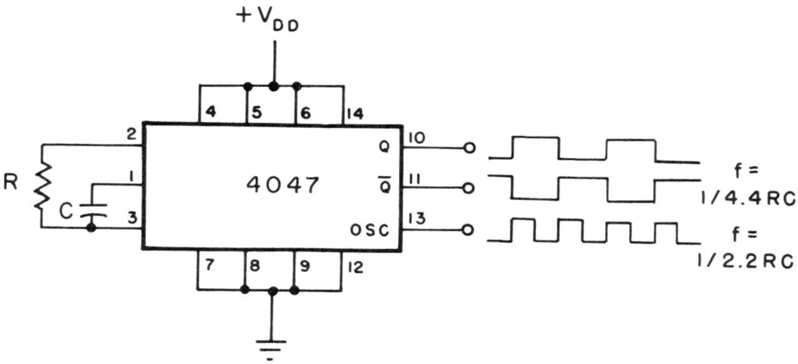

Fig. 5-21. 4047 astable having three outputs.

be obtained at pins 10 and 11, which are complementary outputs. The complementary outputs have a frequency that is one-half that of pin 13. In this case the output frequency (in hertz) is found from

$$f = \frac{1}{4.4RC} \qquad \text{(Eq. 5-15)}$$

For oscillator frequencies up to about 500 kHz at pin 13, the external timing capacitor can be any value greater than 100 pF, while the timing resistor should be in the 10-kΩ to 1-MΩ range.

555 TIMER ASTABLE

Like the monostable multivibrator, the astable 555 timer is, perhaps, the most popular type with circuit designers. Based on the circuit of Fig. 5-22, the output frequency (in hertz) is given by

$$f = \frac{1.443}{(R_1 + 2R_2)C} \qquad \text{(Eq. 5-16)}$$

The percent duty cycle D of the output is determined from resistors R_1 and R_2 so that

$$D = \left(\frac{R_1 + R_2}{R_1 + 2R_2}\right) \times 100 \qquad \text{(Eq. 5-17)}$$

Thus, by making R_2 very large with respect to R_1, the duty cycle approaches 50%. The resistance term in the denominator of Equation 5-16 or Equation 5-17 should be in the 1-kΩ to 10-MΩ range. However, it should be noted that these two timing resistors cannot be adjusted for a given duty cycle without affecting the output frequency of the time. Ae we shall see in the next section, we are able to vary the duty cycle without significant frequency changes.

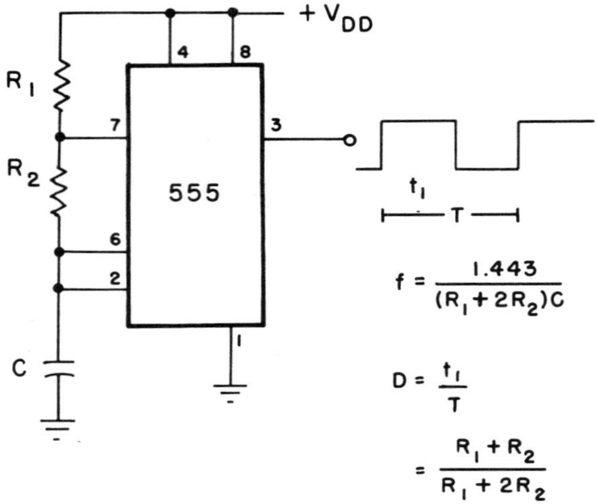

Fig. 5-22. 555 timer astable.

VARIABLE DUTY CYCLE CONSTANT FREQUENCY ASTABLE

Certain applications require that the output waveform from an astable multivibrator have a required symmetry. Such a situation was pointed out in the previous chapter in Fig. 4-14, where the duty cycle of the waveform was the factor controlling the cutoff frequency of an active filter.

Fig. 5-23 shows a variable duty cycle astable circuit for whose range the output frequency is essentially constant. The resistance

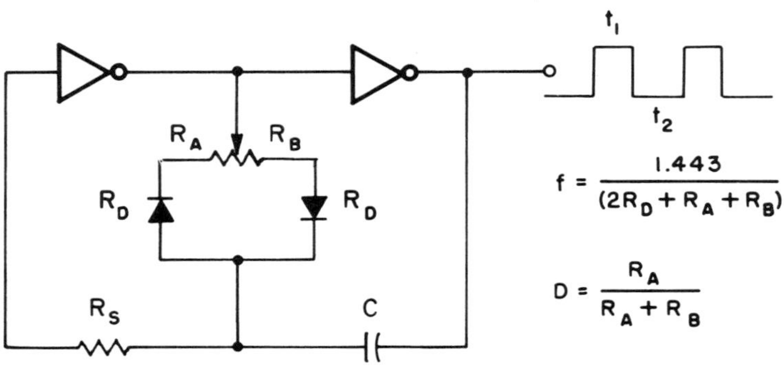

Fig. 5-23. Variable duty cycle, constant frequency control.

of the potentiometer is the sum of resistances R_A and R_B. Resistance R_A plus the forward resistance of diode R_D (typically several hundred ohms) determines the time period that the output is high (t_1), so that, in general,

$$t_1 = -R_A'Cln\left(1 - \frac{V_{TR}}{V_{DD}}\right) \qquad \text{(Eq. 5-18)}$$

where

$$R_A' = R_A + R_D$$

The time period t_2 for which the output is low is determined by R_B and the diode,

$$t_2 = -R_B'Cln\left(\frac{V_{TR}}{V_{DD}}\right) \qquad \text{(Eq. 5-19)}$$

where

$$R_B' = R_B + R_D$$

Both Equations 5-18 and 5-19 can then be ideally simplified to

$$t_1 = 0.693[R_A + R_D]C \qquad \text{(Eq. 5-20)}$$

and

$$t_2 = 0.693[R_B + R_D]C \qquad \text{(Eq. 5-21)}$$

by assuming that V_{TR} is one-half of the supply voltage.

For all practical purposes, the resultant percent duty cycle can be approximated by the relation

$$D = \left(\frac{t_1}{t_1 + t_2}\right) \times 100$$

$$= \left(\frac{R_A}{R_A + R_B}\right) \times 100 \qquad \text{(Eq. 5-22)}$$

if the resistance of the potentiometer ($R_A + R_B$) is considered to be very large when compared to the forward resistance of the diodes. If R_A is less than R_B, the duty cycle will be less than 50%; if greater than R_B, it will be greater than 50%.

The output frequency (in hertz) for such a circuit is then determined by combining Equations 5-20 and 5-21, or

$$f = \frac{1}{t_1 + t_2}$$

$$= \frac{1.443}{(2R_D + R_A + R_B)C} \qquad \text{(Eq. 5-23)}$$

Like the circuit of Fig. 5-18, R_S is made from 2 to 10 times the value of the frequency determining resistance, which in this case is the

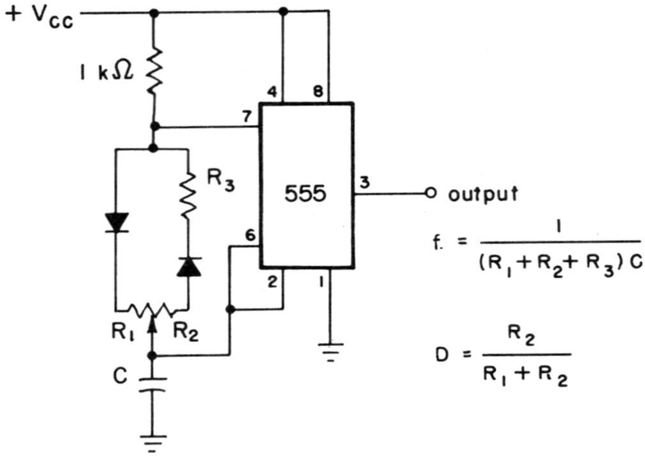

Fig. 5-24. 555 timer variable duty cycle, constant frequency control.

resistance of the potentiometer. Consequently, the sensitivity of the circuit to supply voltage changes is minimized.

Using the 555 timer, another variable duty cycle is shown in Fig. 5-24, which also uses a pair of diodes and a single potentiometer. Since the voltage drop of the diodes decreases the effective voltage across the RC timing network, the output frequency will be slightly higher than that predicted by Equation 5-16. In terms of the timing components, the output frequency (in hertz) is given by

$$f = \frac{1}{(R_1 + R_2 + R_B)C} \qquad \text{(Eq. 5-24)}$$

while the duty cycle (percent) is determined from the relation

$$D = \frac{R_2}{R_1 + R_2} \times 100 \qquad \text{(Eq. 5-25)}$$

VOLTAGE-CONTROLLED OSCILLATORS

Another type of astable multivibrator is the vco, or *voltage-controlled oscillator*. The vco is also called a voltage-to-frequency (V/F) converter, in which the output frequency is proportional to its input control voltage. This section describes circuits that can be constructed using either a 4007 or the monolithic 4046 CMOS phase-locked loop.

In addition to functioning as a transmission gate, the 4007 chip can also be wired as a vco, as shown by the basic circuits of Fig. 5-25, which is a modification of Fig. 5-18. Once the input control

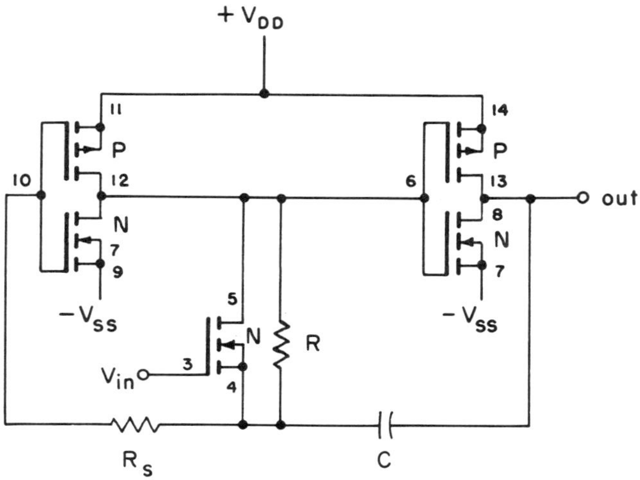

(A) Basic circuit.

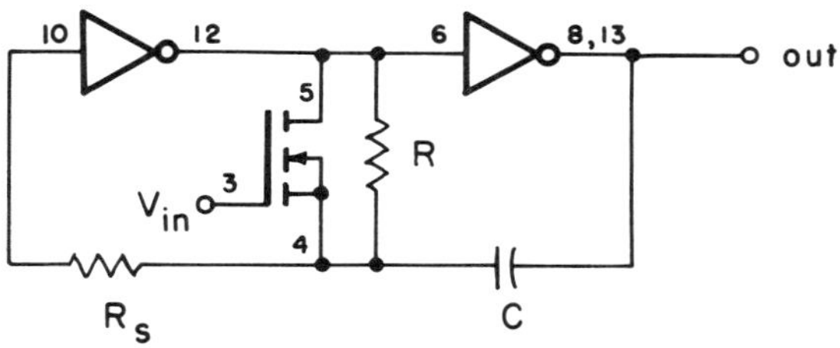

(B) Equivalent circuit.

Fig. 5-25. 4007 voltage-controlled oscillator.

voltage exceeds approximately 1 volt, the drain-to-source resistance R_{DS} of the n-channel MOS transistor varies inversely with the control voltage. The timing resistance of the oscillator is the parallel combination of the drain-to-source resistance and the external resistance R.

When the control voltage is zero, the MOS transistor is off, and R_{DS} is approximately 10^{12} Ω. At the other extreme, R_{DS} is approximately 1 kΩ when the control voltage equals the supply voltage. Consequently, the parallel combination of (R_{eq}) of R_{DS} and R will vary from R to approximately 1 kΩ as the control voltage increases from zero to the supply voltage. The output frequency is essentially that of Equation 5-11, or

$$f = \frac{0.455}{R_{eq}C} \qquad \text{(Eq. 5-26)}$$

where

$$R_{eq} = \frac{R_{DS}R}{R_{DS} + R}$$

The output frequency, therefore, increases with increasing control voltage.

Part of the 4046 CMOS monolithic phase-locked loop can be wired to function solely as a vco*, as shown in Fig. 5-26. The output frequency (in hertz) is determined solely by the equation

$$f = \frac{1}{RC} \qquad \text{(Eq. 5-27)}$$

When the control voltage is zero, the output frequency is zero; when equal to the supply voltage, it equals that given by Equation 5-27.

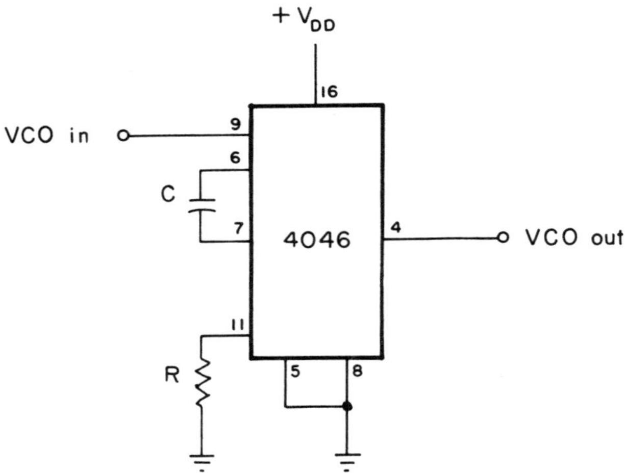

Fig. 5-26. 4046 voltage-controlled oscillator.

The timing resistor must be in the 5-kΩ to 1-MΩ range, while the capacitor can be any value greater than 100 pF. The maximum output frequency is typically 1.5 MHz. When used as a vco, the connections to the phase detector of the 4046, and an external loop filter are not required. Applications using the 4046 are illustrated in Chapter 8.

---

* Additional details about the 4046 device and other aspects of the phase-locked loop are found in *Design of Phase-Locked Loop Circuits, with Experiments,* published by Howard W. Sams & Co., Inc., 1978.

AN INTRODUCTION TO THE EXPERIMENTS

The following experiments are designed to demonstrate the operation of CMOS devices which can be wired as monostable and astable multivibrators, voltage-controlled oscillators, and fixed frequency astables with variable duty cycle. The experiments that you will perform can be summarized as follows:

Experiment No.	*Purpose*
1	Demonstrates the operation of the 4047 multivibrator connected as a monostable multivibrator, which can be triggered by positive or negative going pulses.
2	Demonstrates the operation of the 555 timer as a monostable multivibrator.
3	Demonstrates the operation of an astable multivibrator using either a 40106 or 74C14 Schmitt trigger with a single time constant.
4	Demonstrates the operation of the 555 timer as an astable multivibrator.
5	Demonstrates how the 555 timer can be wired to produce a variable duty cycle astable multivibrator, whose output frequency is relatively constant.
6	Demonstrates the operation of the 4047 multivibrator connected as an astable multivibrator.

EXPERIMENT 1

Purpose

The purpose of this experiment is to demonstrate the operation of the 4047 integrated circuit connected as a monostable multivibrator, which may be triggered by positive or negative-going pulses.

Design Basics

Output pulse width: $\tau = 2.48(RC)$ seconds

for the component ranges: $10\ k\Omega \leqslant R \leqslant 1\ M\Omega$

$C \geqslant .001\ \mu F$

Step 1

Wire the circuit as shown in the schematic diagram, using a 220-kΩ resistor for R and a 10-μF capacitor for C. Then apply power to the breadboard.

Pin Configuration of Integrated Circuit Chip (Fig. 5-27)

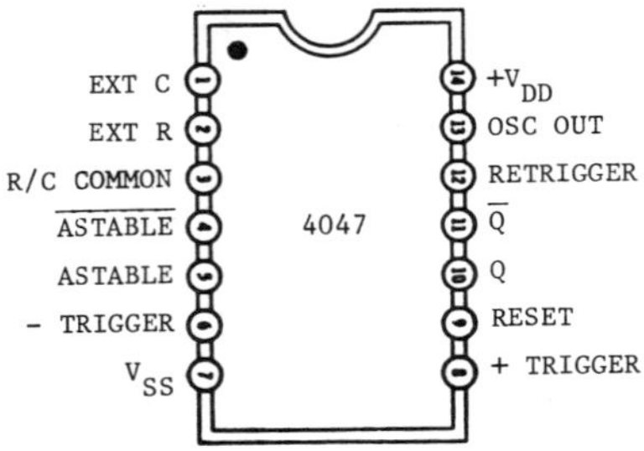

Fig. 5-27.

Schematic Diagram of Circuit (Fig. 5-28)

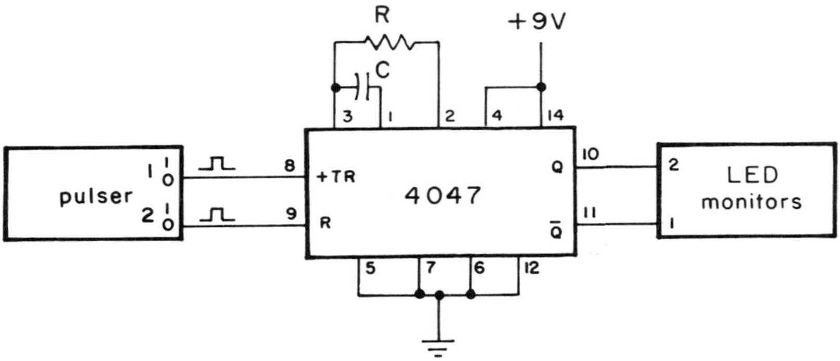

Fig. 5-28.

Step 2

Using either a stopwatch or a wristwatch with a second hand, determine how many seconds LED monitor No. 2 stays lit when pulser No. 1 is quickly pressed and released. Initially you should observe that LED monitor No. 1 is lit and LED monitor No. 2 is unlit. When pulser No. 1 is pressed and released, LED monitor No. 1 immediately goes out while LED monitor No. 2 lights up. The multivibrator is now *triggered.* You should have determined that LED monitor No. 1 stays lit for approximately 6 seconds.

This circuit is triggered by positive trigger pulses, since it responds to *negative-to-positive* transitions of the trigger pulse. For this reason, the trigger pulse is connected to the +TRIGGER input (pin 8) of the 4047.

Step 3

Disconnect the power from the breadboard. Replace the 220 kΩ timing resistor with a 390 kΩ resistor, and then reconnect the power to the breadboard. Again, quickly press and release pulser No. 1 and determine the time that LED monitor No. 2 is lit.

You should have measured approximately 10 seconds. Confirm this with the equation given in the "Design Basics" section.

Step 4

Again, quickly press and release pulser No. 1. Then press and release pulser No. 2, which is connected to the RESET input of the 4047. What happens to the LED monitors?

When pulser No. 2 is pressed, LED monitor No. 2 goes out while LED monitor No. 1 lights up. When the RESET input changes from logic 0 to logic 1, the timing cycle is terminated, so that the Q output (LED No. 2) returns to logic 0 (unlit) while the Q output returns to logic 1, since Q and $\overline{Q}$ are *complementary outputs.* At any time during the timing cycle, the cycle can be terminated by setting the RESET input at logic 1.

Step 5

Disconnect the power from the breadboard. Now connect pin 8 to the positive supply voltage V_{DD}, and connect pin 6 (the −TRIGGER input) to the output of pulser No. 1 that is normally at logic 1. This circuit will now be triggered by *negative* trigger pulses, since the circuit now responds only to the *positive-to-negative* transitions of the trigger pulse. Now quickly press and release pulser No. 1 and measure the time that LED monitor No. 2 is lit.

You should find that LED monitor No. 2 remains lit for approximately 10 seconds, just as was determined in Step 3.

Step 6

Repeat Step 4. Is the result the same as was observed in Step 4?

The two results are the same, since the 4047 monostable multivibrator is reset when the RESET input is at logic 1, regardless of positive or negative triggering.

EXPERIMENT 2

Purpose

The purpose of this experiment is to demonstrate the operation of the 555 timer as a monostable multivibrator.

Pin Configuration of Integrated-Circuit Chip (Fig. 5-29)

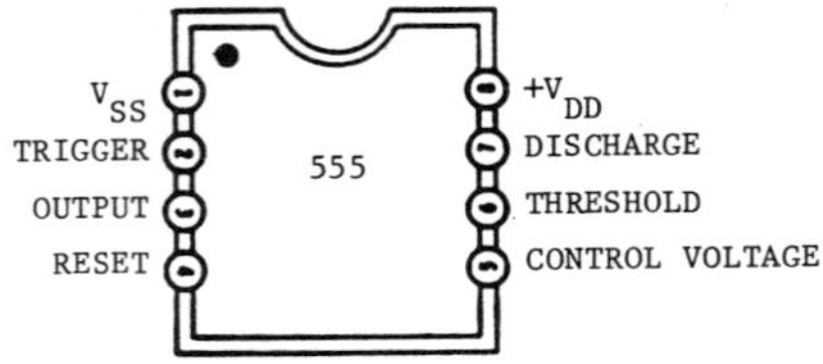

Fig. 5-29.

Schematic Diagram of Circuit (Fig. 5-30)

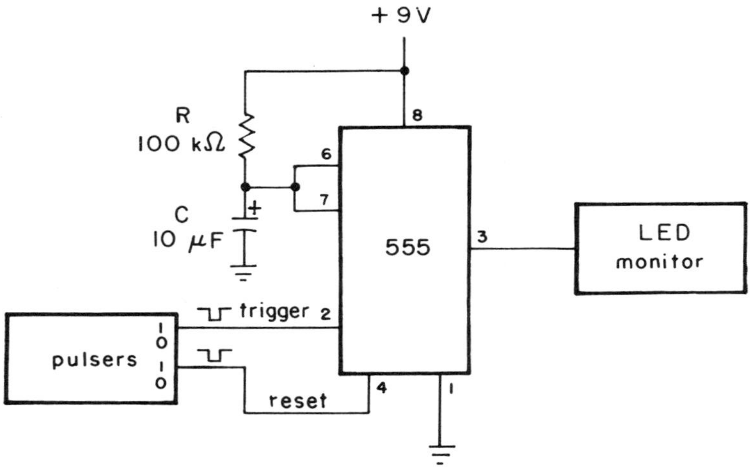

Fig. 5-30.

Design Basics

Output pulse width: $\tau = 1.1(RC)$ seconds

Step 1

Wire the circuit as shown in the schematic diagram, first using a 100-kΩ resistor for R and a 10-μF electrolytic capacitor for C. When using electrolytics, pay careful attention to the polarity! A normally open push-button switch is connected from pin 2 of the timer to ground, or a pulser (debounced switch, see Fig. 2-6) can be used instead.

Step 2

Using either a stopwatch or a wristwatch with a second hand, determine how many seconds the LED monitor stays lit when the pulser or push-button switch is quickly pressed and released. Next, using the various combinations of resistors and capacitors listed below, determine the output pulse width (i.e., the time that the LED is lit). Write your answers below and compare them with the design equation.

Timing Resistor R	Timing Capacitor C	Measured Pulse Width
100 kΩ	10 μF	
100 kΩ	100 μF	
1 MΩ	5 μF	
330 kΩ	10 μF	
330 kΩ	100 μF	
10 MΩ	1 μF	
620 kΩ	5 μF	
10 MΩ	5 μF	

The results that you obtain should be consistent to within 10% when compared with the design equation. If not, chances are that the capacitor may be "leaky," so that it is difficult to charge and, consequently, it requires more time. Therefore, it is important to try to use either Mylar or tantalum type electrolytic capacitors for the timing capacitor when using values greater than 1 microfarad, rather than the commonly available aluminum types.

Step 3

Now use a 330-kΩ resistor for R and a 10-μF capacitor for C. Press the pulser in for approximately 5 seconds while watching the LED monitor.

You should have observed that the LED remained lit until you released the pulser, or, perhaps, the LED remained lit approximately 3.6 seconds after you relased the pulser. The reason is that the time duration of the trigger pulse exceeded the time duration of the output pulse width. Consequently, the negative-going trigger pulse should be kept short compared with the desired output pulse width.

Step 4

Replace the timing resistor R with a 1-MΩ resistor. As before in Step 2, quickly press and release the pulser that is connected to the TRIGGER input (pin 2) of the 555 timer. The LED should remain lit for about 11 seconds. Again, quickly press and release the pulser, and then press and release the pulser that is connected to the RESET pin of the timer (pin 4). You should observe that the LED immediately goes out. From this action, it should be clear that the reset inhibits the timer when pin 4 is changed from a logic 1 to a logic 0 state, forcing the output to return to logic 0.

Consequently, this monostable timer can be reset at any time during the timing cycle simply by grounding the RESET pin of the timer. To trigger the timer again, pin 4 must be at logic 1.

EXPERIMENT 3

Purpose

The purpose of this experiment is to demonstrate the operation of an astable multivibrator using a 40106 (or 74C14) Schmitt trigger with a single time constant.

Pin Configuration of Integrated Circuit Chip (Fig. 5-31)

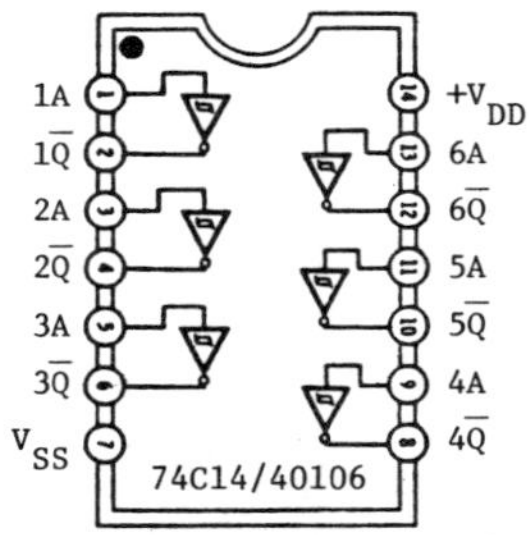

Fig. 5-31.

Schematic Diagram of Circuit (Fig. 5-32)

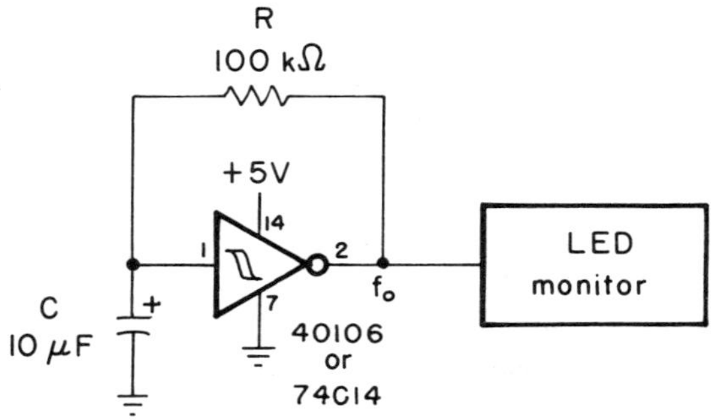

Fig. 5-32.

Design Basics

$$\text{Output frequency: } f = \frac{K}{RC}$$

where the constant κ depends on the device and the manufacturer, as summarized in Table 5-3. In general, the output frequency increases with supply voltage even though the timing component values remain constant.

Step 1

Wire the circuit as shown in the schematic diagram. Since unused inputs of CMOS devices cannot be left floating, connect the remaining five inputs: pins 3, 5, 9, 11, and 13 to ground.

Step 2

Apply power to the breadboard. What happens to the status of the LED monitor?

The LED monitor should be flashing off and on, indicating that the output of the circuit is alternating between logic 0 (ground) and logic 1 (+5 volts), which is an astable multivibrator.

Step 3

Disconnect the power from the breadboard and replace the 10-μF capacitor with a 0.01-μF capacitor. Apply power to the breadboard and measure the output frequency with a frequency counter. If you are using either a 74C14 or a 40106 Schmitt trigger made by National Semiconductor, the constant κ in the design equation is typically 0.529 for a 5-volt supply; if the 40106 is made by RCA, it is typically 1.231. How does the measured output frequency compare with the design equation?

Within 15%, you should have measured 529 Hz (for National Semiconductor chips), or 1231 Hz for the 40106 made by RCA. If you are using a 40106 Schmitt trigger made by RCA, skip the next step and continue with Step 5.

Step 4

Now increase the power supply voltage to 10 volts and measure the output frequency. Is there any change?

You should measure an increase in frequency, assuming that you are using either the 74C14 or 40106 devices made by National Semiconductor. The proportionality constant κ in the design equation for these chips takes into account the lower and upper threshold voltages of the Schmitt trigger, which are dependent on the supply voltage. At 10 volts, κ is typically 0.663, so that your measured output frequency should be within 15% of 663 Hz.

Step 5

Disconnect the power from the breadboard. Replace the 100-kΩ resistor with a 12-kΩ resistor, and the 0.01-μF capacitor with a 0.0022-μF capacitor. Connect the power to the breadboard and repeat Steps 3 and 4.

For the National 74C14 and 40106 devices, the measured output frequency with a supply voltage of 5 volts should be about 20.0 kHz; with the 40106 by RCA, it should be about 46.7 kHz.

When the supply voltage is increased to 10 volts, the output frequency for either of the devices made by National Semiconductor should be about 25.1 kHz; for the RCA device, there should be virtually no change, since its triggering threshold increases proportionately with the supply voltage.

EXPERIMENT 4

Purpose

The purpose of this experiment is to demonstrate the operation of the 555 timer connected as an astable multivibrator.

Pin Configuration of Integrated Circuit Chip (Fig. 5-33)

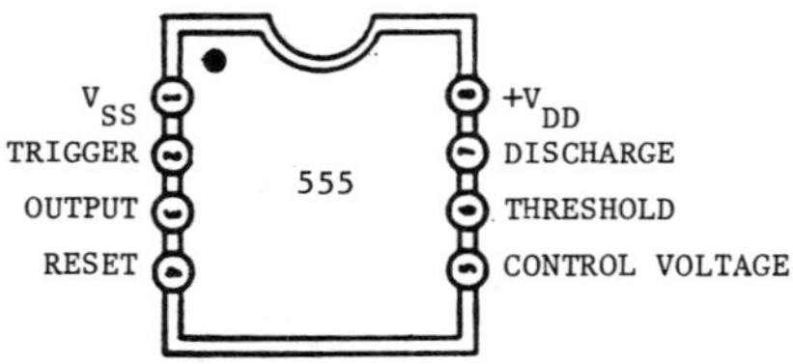

Fig. 5-33.

Design Basics

$$\text{Output frequency (in hertz):}\quad f = \frac{1.443}{(R_1 + 2R_2)C}$$

$$\text{Percent duty cycle:}\quad D = \frac{R_1 + R_2}{R_1 + 2R_2}(100)$$

Step 1

Wire the circuit as shown in the schematic diagram using a 1-kΩ resistor for R_1, a 1-MΩ resistor for R_2, and a 1-μF capacitor for C.

Schematic Diagram of Circuit (Fig. 5-34)

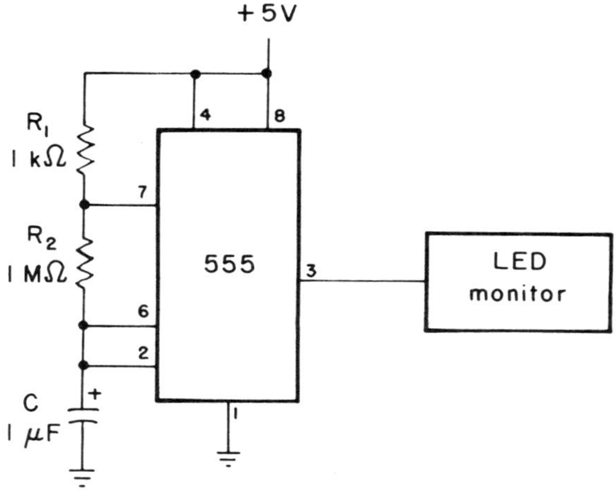

Fig. 5-34.

Step 2

Connect the power to the breadboard and observe that the LED monitor periodically flashes on and off. With the aid of a stopwatch or a wristwatch having a second hand, count the number of times the LED monitor is lit over a 1-minute period.

You should count about 44 pulses per minute, which is the same as a frequency of 0.73 Hz and is slow enough for the eye to follow.

Step 3

Disconnect the power from the breadboard. Change the capacitor value to 0.001 μF. Based on the resistance values for R_2 listed below, measure the output frequency of the timer and compare the measured frequency with the values predicted by the equation given in the "Design Basics" section.

Timing Resistor R_2	Calculated Frequency	Measured Frequency
330 kΩ	4373 Hz	
100 kΩ		
15 kΩ		
10 kΩ		
4.7 kΩ		

Step 4

Disconnect the power from the breadboard. Now use a 100-kΩ resistor for R_2 and a 0.01-μF capacitor for C. In addition, connect an oscilloscope to the output (pin 3) of the 555 timer.

Step 5

Apply power to the breadboard. What is the output frequency and duty cycle of the square wave?

You should have determined that the output frequency is about 720 Hz. The duty cycle is the ratio of the time period that the output is HIGH, or +5 volts, to the time for one complete cycle. In this case, you should find that the duty cycle is very nearly 50%. Compare this with the equation given in the "Design Basics" section.

Step 6

Disconnect the power from the breadboard and replace the 1-kΩ resistor (R_1) with a 100-kΩ resistor, and the 100-kΩ resistor (R_2) with a 47-kΩ resistor. Connect the power to the breadboard. What is the output frequency and duty cycle now?

Although the output frequency increases slightly (to about 740 Hz), you should have measured a duty cycle of about 75%, which is a 50% increase for less than a 5% increase in frequency.

Step 7

Try different resistor combinations for R_1 and R_2 to try to produce a wide range of duty cycles. In this experiment you should conclude that it is difficult to vary the duty cycle of the output waveform while keeping the output frequency constant, using the circuit of Fig. 5-34. A more practical circuit is described in the next experiment.

EXPERIMENT 5

Purpose

The purpose of this experiment is to demonstrate that it is possible, using a 555 timer, to vary the duty cycle of an astable multivibrator while holding the output frequency relatively constant.

Pin Configuration of Integrated Circuit Chip (Fig. 5-35)

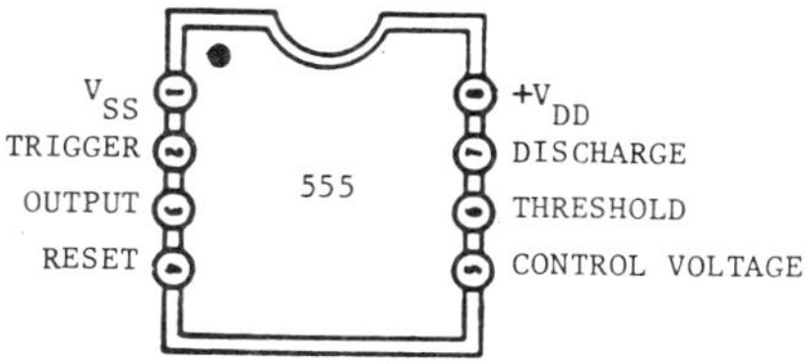

Fig. 5-35.

Schematic Diagram of Circuit (Fig. 5-36)

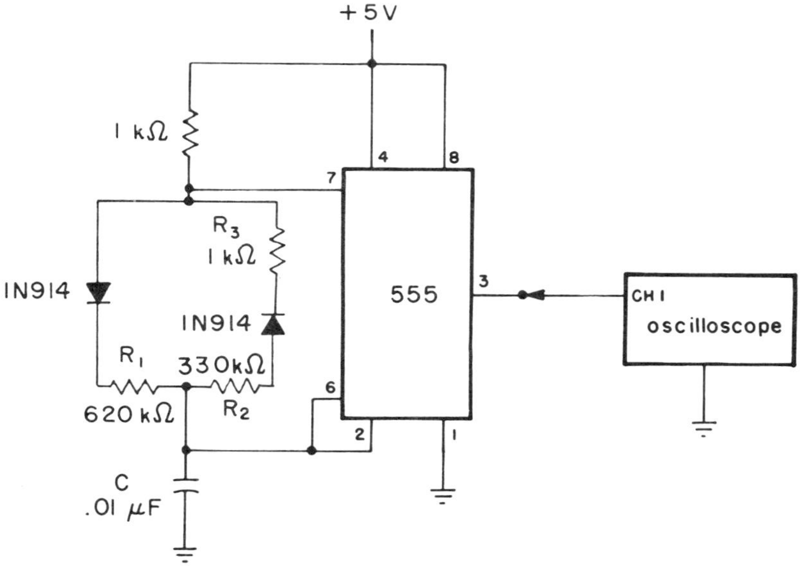

Fig. 5-36.

Design Basics

$$\text{Output frequency (in hertz):} \quad f = \frac{1}{(R_1 + R_2 + R_3)C}$$

$$\text{Percent duty cycle:} \quad D = \left(\frac{R_2}{R_1 + R_2}\right) \times 100$$

Step 1

Wire the circuit as shown in the schematic diagram, paying careful attention to the polarity of the two diodes.

Step 2

Apply power to the breadboard and measure the output frequency and duty cycle.

You should measure an output frequency that is approximately 105 Hz. The time period that the output is at logic 1 should be approximately 3.3 ms, so that with a total cycle period of 9.5 ms, the duty cycle is approximately 35%. Compare these results with the design equations.

Step 3

Briefly disconnect the power from the breadboard and interchange the 330-kΩ and 620-kΩ resistors. The 330-kΩ resistor is now R_1, while R_2 is now 620 kΩ. Then reconnect the power to the breadboard and measure the output frequency and duty cycle. What, if anything, happened?

You should have observed that the duty cycle increased from 35% to 65%, but there was *no change in the output frequency!* The duty cycle, consequently, depends on the position of the *larger* resistance (620 kΩ). If R_2 is larger than R_1, the duty cycle will be greater than 50%. Also, if R_2 is less than R_1, the duty cycle will be less than 50%.

Step 4

Disconnect the power from the breadboard and replace the 330-kΩ and 620-kΩ resistors with a 1-MΩ potentiometer, as shown in Fig. 5-37.

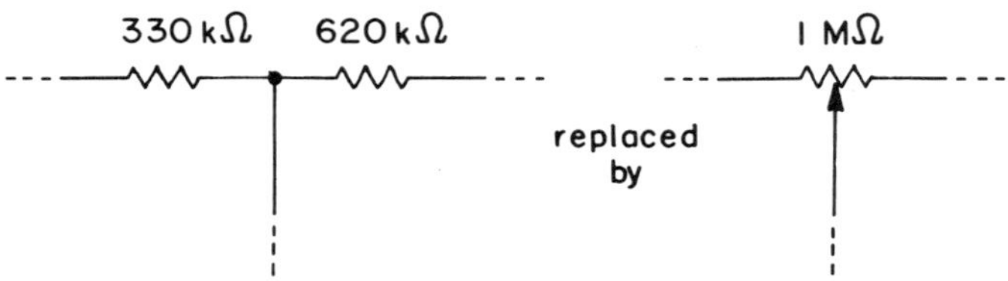

Fig. 5-37.

Step 5

Reconnect the power to the breadboard. Now vary the resistance of the potentiometer from one extreme to the other. What do you observe about the duty cycle and output frequency as the potentiometer resistance is varied?

You should have noticed that the duty cycle varied from almost 3% to about 97% without any change in frequency, depending on the setting of the potentiometer. This circuit is very useful for smoothly changing the duty cycle of a 555 timer astable multivibrator with only a single adjustment, which does not affect the output frequency.

EXPERIMENT 6

Purpose

The purpose of this experiment is to demonstrate the operation of the 4047 integrated circuit connected as an astable multivibrator.

Pin Configuration of Integrated Circuit Chip (Fig. 5-38)

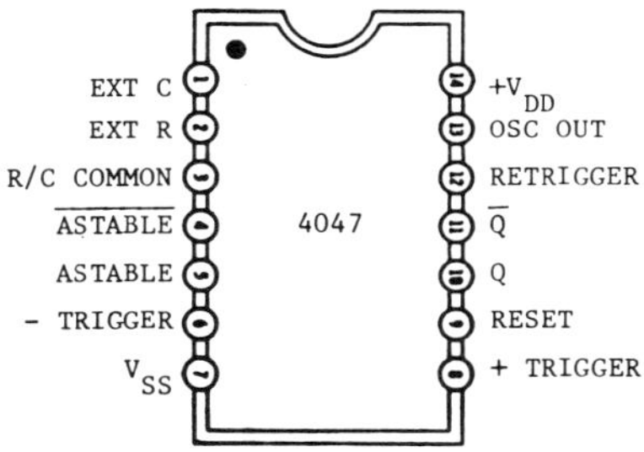

Fig. 5-38.

Design Basics

Output frequency (in hertz): $f = \dfrac{1}{2.2(RC)}$ @ pin 13

$f = \dfrac{1}{4.4(RC)}$ @ pin 10 or 11

Step 1

Wire the circuit as shown in the schematic diagram and apply power to the breadboard.

Step 2

With the aid of a stopwatch or a wristwatch having a second hand, count the number of times LED monitor No. 1 flashes on and off over a 1-minute period.

Schematic Diagram of Circuit (Fig. 5-39)

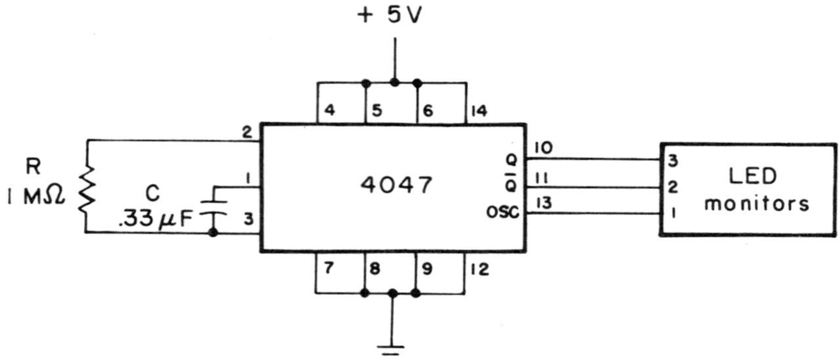

Fig. 5-39.

Depending upon the accuracy of your timing components, you should count about 82 flashes per minute, which is the same as a frequency of 1.38 Hz and is slow enough for the eye to follow.

Step 3

Now count the number of times LED monitors No. 2 and No. 3 each flash on and off over a 1-minute period. How does this compare with what you determined in Step 2?

You should have observed that LED monitors No. 2 and No. 3 flash at a rate that is one-half the rate you measured from LED monitor No. 1. Do you see any relationship between LED monitors No. 2 and No. 3?

You should observe that when LED monitor No. 2 is off, LED monitor No. 3 is on, and vice versa. Consequently, these two LED monitors indicate that the two outputs at pins 10 and 11 are complementary outputs, which both have an output frequency that is one-half that of the OSCILLATOR OUTPUT of the 4047 (pin 13).

Step 4

Now disconnect the power from the breadboard and replace the 0.33-μF capacitor with a 0.01-μF capacitor, and replace the 1-MΩ resistor with a 10-kΩ resistor. Reconnect the power to the breadboard and connect a frequency counter to pin 13 of the 4047 integrated circuit. Measure the output frequency. How does it compare with the design equation?

Depending on the accuracy of your timing components, the OSCILLATOR frequency at pin 13 should be approximately 4545 Hz. Now measure the frequency at pins 10 and 11. How does it compare with the frequency at pin 13?

The frequency at pins 10 and 11 should be exactly one-half that of pin 13, or approximately 2273 Hz. Your reading will probably differ by ±1 Hz since the frequency counter has an error of ±1 Hz. Depending on a particular application, the 4047 astable multivibrator is capable of generating two output frequencies, one that is one-half the other.

CHAPTER 6

Counters

INTRODUCTION AND OBJECTIVES

Digital counters are one of the most important logic elements in the arsenal of CMOS integrated circuits. Both binary ripple and synchronous counters are present in the timing and display of many digital instruments.

This chapter discusses both the binary ripple and synchronous CMOS circuits, which are primarily used as divide-by-*N* counters. However, several types have the ability to be wired to synthesize various periodic waveforms or frequencies digitally. Finally, binary rate multipliers can be used to perform a number of arithmetic operations, such as addition, subtraction, multiplication, division, ratios, and roots.

At the conclusion of this chapter, you will be able to do the following:

- Explain the differences between binary ripple and synchronous counters.
- Describe the operation of the following synchronous counter types:
    1. Counters with decoded 1 of *N* outputs, such as the 4017 and 4022.
    2. Decade counters with 7-segment decoded outputs, such as the 4026 and 4033.
    3. Dual up BCD and binary counters, such as the 4518 and 4520.
    4. Programmable up-down BCD and binary counters, such as the 4029, 4510, 4516, 40192, and 40193.

5. Programmable down BCD and binary counters, such as the 4059, 40102, and 40103.
6. Multidecade programmable counters such as the Hughes 0320.

- Describe how the 4060 binary ripple counter can be used with a crystal to form an oscillator and binary divider.
- Describe how binary ripple and synchronous counters can be used to synthesize certain periodic waveforms.
- Describe the operation of BCD and binary rate multipliers, such as the 4089 and 4527 to synthesize arithmetic functions.

BINARY RIPPLE COUNTERS

Binary ripple counters are basically composed of a series of cascaded master-slave flip-flops, so that the output of one counter stage is the clocking input for the next stage, as illustrated in Fig. 6-1 for a simple N-stage binary ripple counter. Output Q_1 of the first stage has a frequency that is one-half of the input clock, while output Q_3 is $\frac{1}{2}^3$ or ⅛ of the input frequency of the counter. From beginning to end, the output of each successive stage is one-half of the preceding stage, so that the modulus increases by powers of 2: 2^1, 2^2, 2^3, etc.

Because the output of any stage is the input clock of the next stage, propagation delay from input to output can be a problem, especially if the counter has a large number of stages, such as the 4045, which is a 21-stage counter. Table 6-1 lists the binary ripple 4000 series CMOS counters available. Other ripple counters, such as the 74C90 (÷10) and 74C93 (÷16) are identical CMOS pin-for-pin equivalents of the corresponding 7490 and 7493 TTL counters.

Of those listed in Table 6-1, the 4045 and 4060 counters can be used with an external crystal to form a stable oscillator. As shown in Fig. 6-2, the 4060 is a combination crystal-controlled oscillator

Table 6-1. CMOS Binary Ripple Counters

			Max Input Freq, MHz			
Device	**Stages**	**Clocking**	**5 V**	**10 V**	**15 V**	**Remarks**
4020	14	┐_	7	16	24	no output for 2^2, 2^3
4024	7	┐_	7	16	24	
4040	12	┐_	7	16	24	
4045	21	┐_	7	16	24	can be used as oscillator with crystal; 2^{21} output only
4060	14	┐_	7	16	24	can be used as oscillator with crystal; no 2^2, 2^3, or 2^{11} output

and 14-stage binary divider. In addition to the fundamental output frequency f_0, determined by the resonant frequency of the crystal, the 4060 also divides this frequency in binary multiples from 16 to

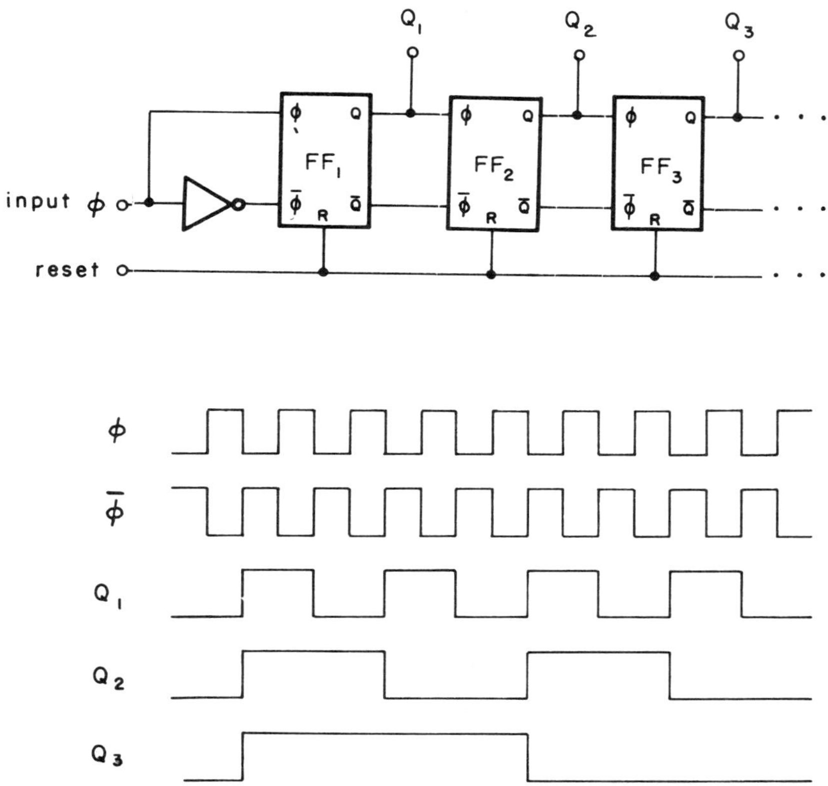

Fig. 6-1. N-stage binary ripple counter.

16,384, except for 2^{11}, or 2048, so that there are a total of 11 different possible output frequencies.

SYNCHRONOUS COUNTERS

Synchronous counters differ from binary ripple counters in that all the internal counter stages are driven from a single clocked input, and all the outputs change simultaneously. As summarized in Table 6-2, CMOS synchronous counters generally fall into one of the following classes:

- Decoded 1-of-*N* outputs
- Decade counters with 7-segment decoded outputs
- Fixed and programmable decade and binary (hexadecimal) up counters

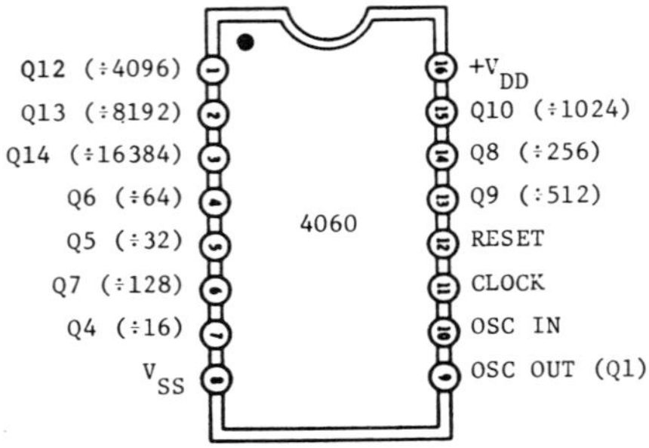

(A) Pin diagram.

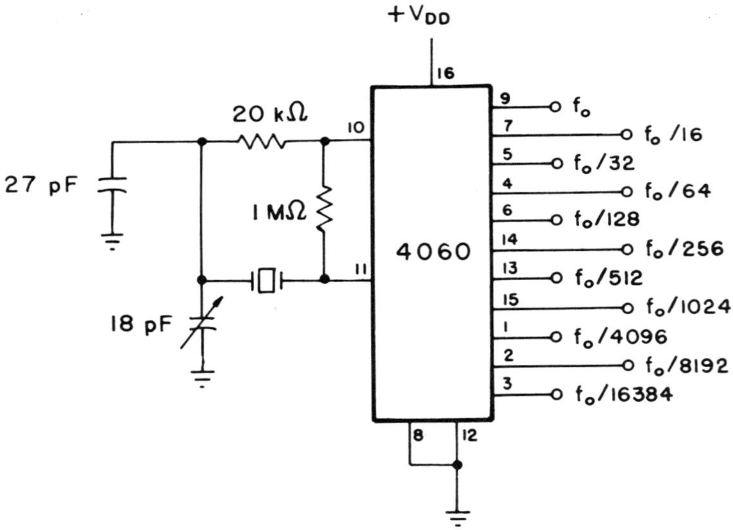

(B) Crystal-controlled oscillator with 11 output frequencies.

Fig. 6-2. 4060 14-stage binary ripple counter/divider and oscillator.

- Programmable up-down decade and binary (hexadecimal) counters
- Programmable down decade and binary (hexadecimal) counters

which are discussed in detail in the following sections.

Table 6-2. CMOS Synchronous Counters

Device	Type	Modulus	Clocking	Max Input Freq, MHz			Remarks
				5 V	10 V	15 V	
0320	programmable	3 to 1024	_|‾	5	10	—*	binary & BCD programming
4017	decoded 1-of-10	10	_|‾	5	12	16	5-stage Johnson counter
4018	programmable	2 to 10	_|‾	5	10	15	Johnson counter
4022	decoded 1-of-8	8	_|‾	5	12	16	4-stage Johnson counter
4026	decade w/7-segment	10	_|‾	3	6	7.5	5-stage Johnson counter; display enable input
4029	programmable up/down	10 or 16	_|‾	4.6	10	14	4-bit binary or BCD programming
4033	decade w/7-segment	10	_|‾	3	6	7.5	5-stage Johnson counter; ripple blanking I/O
4059	programmable down	3 to 9999 or 15,999	_|‾	3	6	—†	4-BCD sections
4510	programmable up/down	10	_|‾	4	8	11	4-bit BCD
4516	programmable up/down	16	_|‾	4	8	11	4-bit binary (hexadecimal)
4518	up	10	_|‾ or ‾|_	3	6	8	dual decade sections
4520	up	16	_|‾ or ‾|_	3	6	8	dual binary sections
40102	programmable down	0 to 100	_|‾	1.4	3.6	4.8	2-decade BCD inputs
40103	programmable down	0 to 256	_|‾	1.4	3.6	4.8	8-bit binary inputs
40192 74C192	programmable up/down	0 to 9	_|‾ or ‾|_	4	8	11	4-bit BCD
40193 74C193	programmable up/down	0 to 15	_|‾ or ‾|_	4	8	11	4-bit binary

* Not specified; maximum V_{DD} is 13 V.
† Not specified; maximum V_{DD} is 12 V.

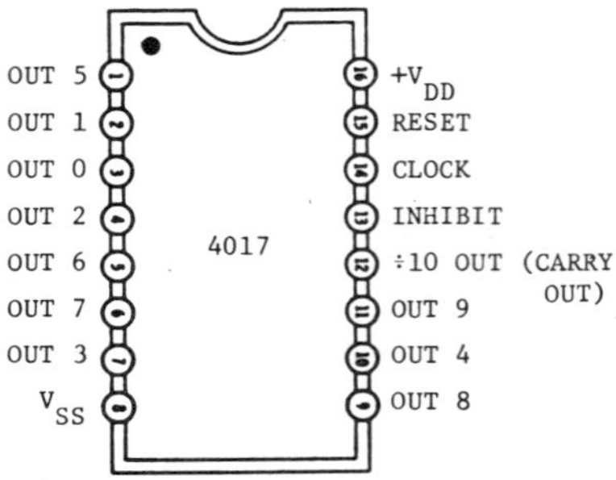

(A) 4017 (1-of-10) pin diagram.

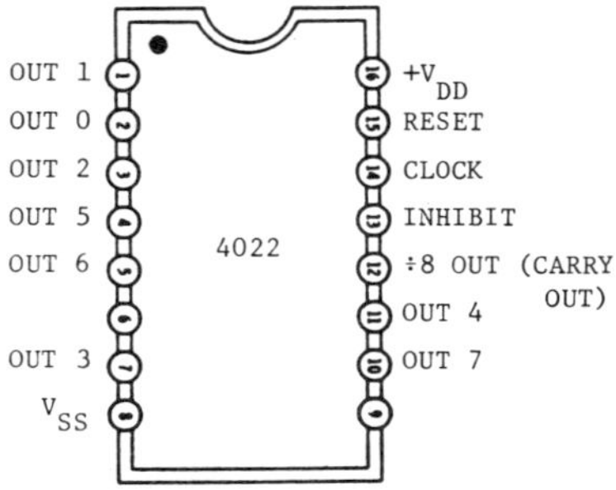

(B) 4022 (1-of-8) pin diagram.

Fig. 6-3. Decoded 1-of-N counters.

Decoded 1-of-N Counters

Decoded 1-of-*N* counters such as the 4017 and 4022 (Fig. 6-3) are unique devices in that they both possess an internal walking ring, or *Johnson* counter with decoded outputs. The 4017 is a decade (÷10) counter with 10 decoded outputs, while the 4022 is a divide-by-8 counter with 8 decoded outputs. These are used primarily in digital filters and waveform generators. Both counters are clocked on the positive leading edge of the input if the INHIBIT input is at logic 0. Otherwise, counting is inhibited if the INHIBIT input is at logic 1. Both counters are reset to the zero count when the RESET input is at logic 1. The decoded outputs are normally low and go high only at their respective time slot, as shown in Fig. 6-4.

Decade Counters With 7-Segment Decoded Outputs

Decade counters, such as the 4026 and 4033 consist of a 5-stage Johnson counter with an output decoder which converts the walking-ring code to decoded 7-segment outputs suitable for driving *common-cathode* LED displays (Chapter 7 discusses specific interfacing techniques for LED displays). The 4026 (Fig. 6-5) has a DISPLAY ENABLE capability, which is used to turn off the display when the DISPLAY ENABLE input is at logic 0. This allows for brightness control using a variable duty cycle astable multivibrator, or to conserve power.

The CARRY OUT (CO) signal is a divide-by-10 output, but is not inhibited by the state of the DISPLAY ENABLE input, which can be used continuously to directly clock the succeeding decade in a multidecade display system.

The 4033 (Fig. 6-6), on the other hand, does not have a DISPLAY ENABLE input, but has provision for automatic suppression, or blanking of the nonsignificant zeros in a multidigit decimal num-

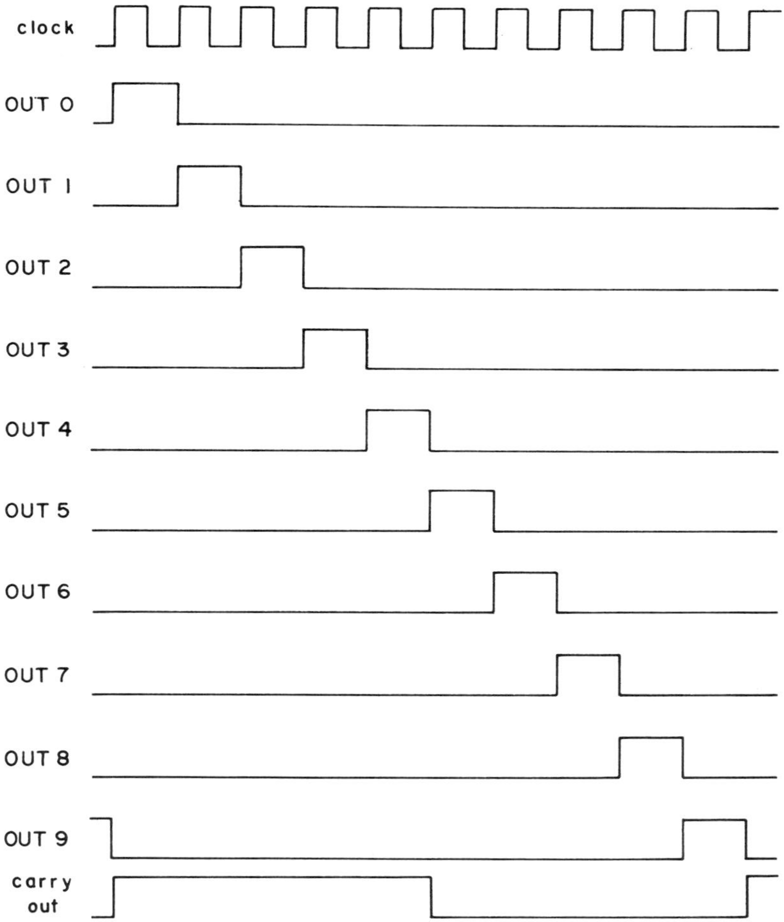

Fig. 6-4. 4017 timing diagram.

ber display. For example, an unblanked 8-digit display may read 0054.3210; with blanking, the nonsignificant zeros are eliminated, giving 54.321. Zero suppression for zeros to the left of the decimal point is accomplished by connecting the RIPPLE BLANKING INPUT (RBI) of the most significant 4033 decade of the display to logic 0, while the RIPPLE BLANKING OUTPUT (RBO) of this stage is connected to the RBI pin of the next-lower significant 4033 decade, as illustrated in Fig. 6-7.

For digits to the right of the decimal point, the RBI pin of the least significant decade is tied to logic 0, while the RBO pin of this

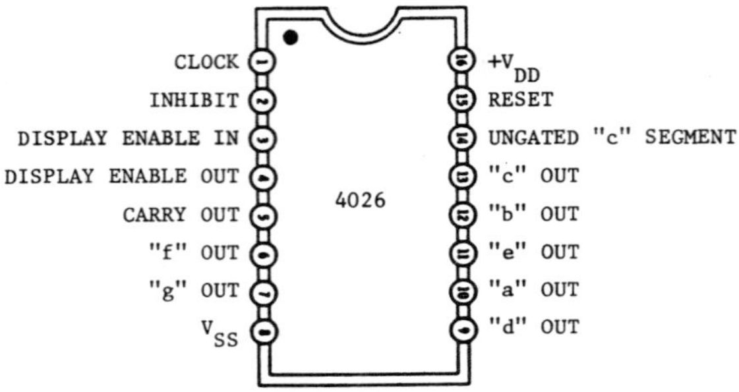

Fig. 6-5. Pin diagram of 4026 decade counter/divider.

4033 decade is connected to the RBI pin of the next-more significant decade, and so on.

Fig. 6-8 shows the general scheme for a multidecade counter/display using either the 4026 or 4033. For more than 4 decades, it is more efficient to multiplex the displays. Chapter 7 discusses how the 4026 ond 4033 are interfaced with multiplexed displays.

Dual Up Counters

The 4518 dual BCD and 4520 dual binary or hexadecimal synchronous up counters (Fig. 6-9) both consist of a pair of identical 4-stage counters which have interchangeable CLOCK and ENABLE inputs. This feature allows either counter to be clocked on positive or negative transitions, as shown in Fig. 6-10 and summarized in Table 6-3.

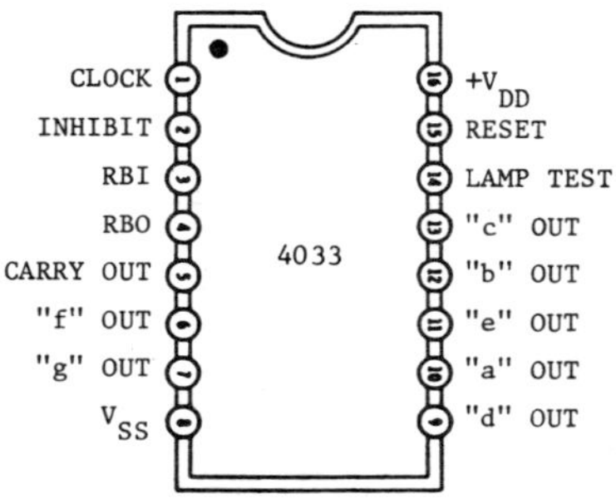

Fig. 6-6. Pin diagram of 4033 decade counter/divider.

Both counters can be cascaded in the *ripple* mode by connecting output Q_4 to the ENABLE input of the next counter, as shown in Fig. 6-11. As with other ripple counters, such as the 4020, propagation delay increases for each counter added. Fig. 6-12 shows how both counters can be synchronously cascaded with an AND gate.

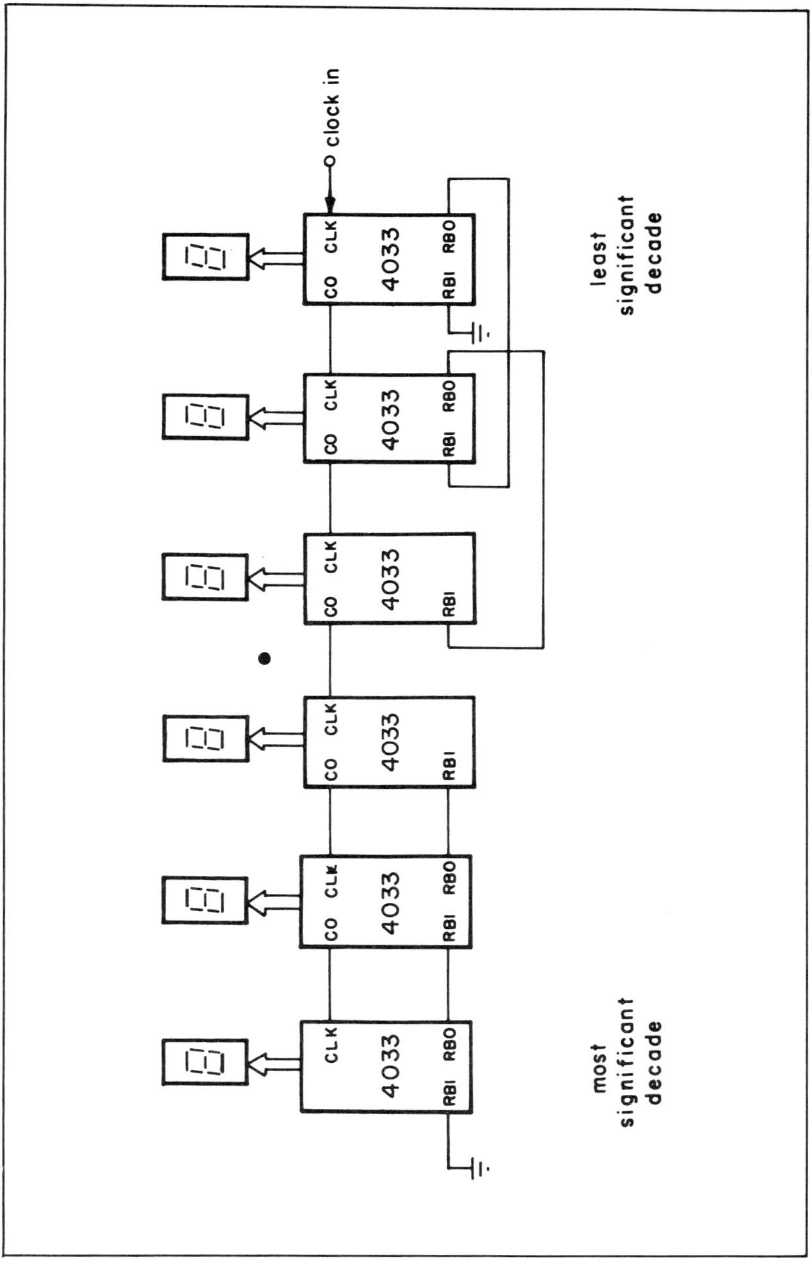

Fig. 6-7. Leading zero suppression connections.

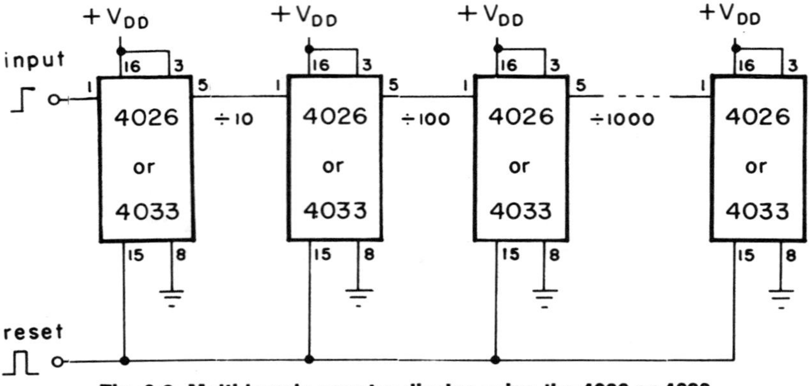

Fig. 6-8. Multidecade counter display using the 4026 or 4033.

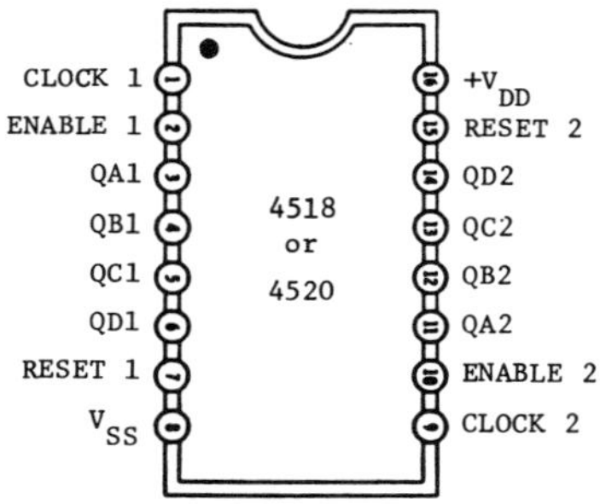

Fig. 6-9. Pin diagram of the 4518 (decade)/4520 (hexadecimal) counters.

When combined with a 4511 BCD/7-segment latch decoder/driver, as shown in Fig. 6-13, we have an arrangement equivalent to the 7490/7475/7448 sequence of TTL devices. Chapter 7 discusses the interfacing of the 4511 decoder with 7-segment LED displays.

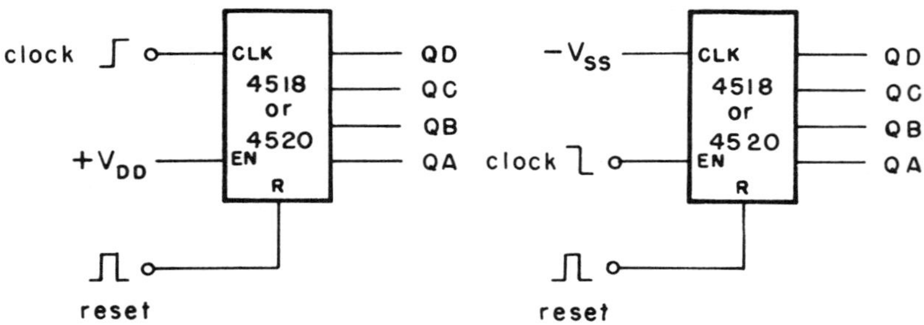

(A) Positive edge clocking.

(B) Negative edge clocking.

Fig. 6-10. 4518 and 4520 counters.

Table 6-3. 4518/4520 Truth Table

CLOCK	ENABLE	Result
↑	1	count up
0	↓	count up
↓	0 or 1	no change
0 or 1	↑	no change
↑	0	no change
1	↓	no change

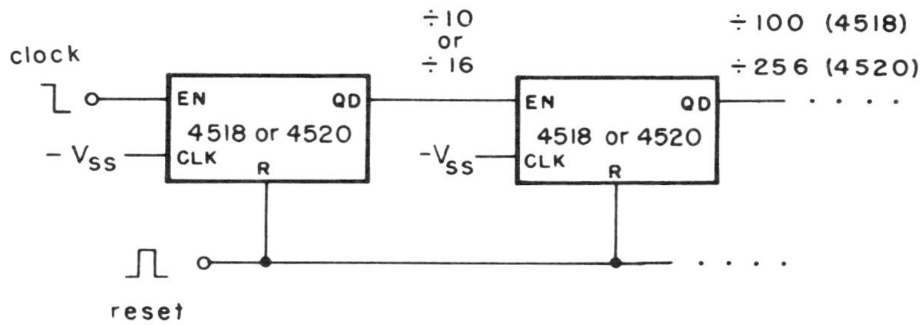

Fig. 6-11. Ripple cascading of 4518/4520 counters.

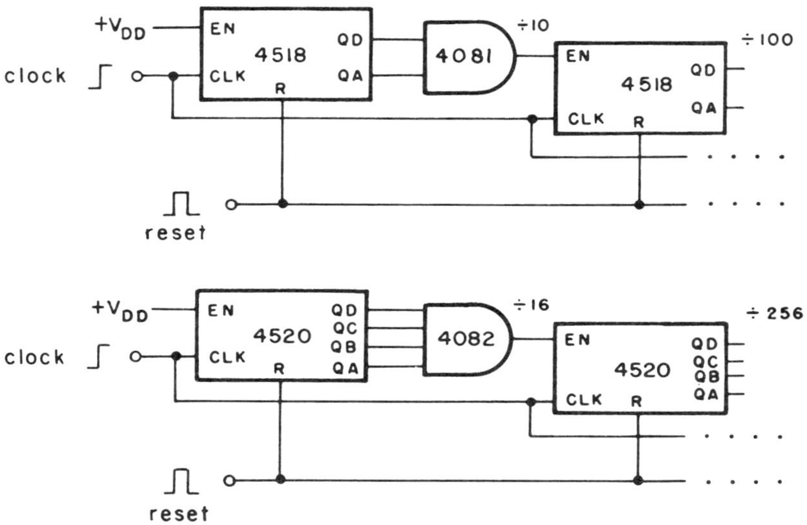

Fig. 6-12. Synchronous cascading of 4518/4520 counters with AND gates.

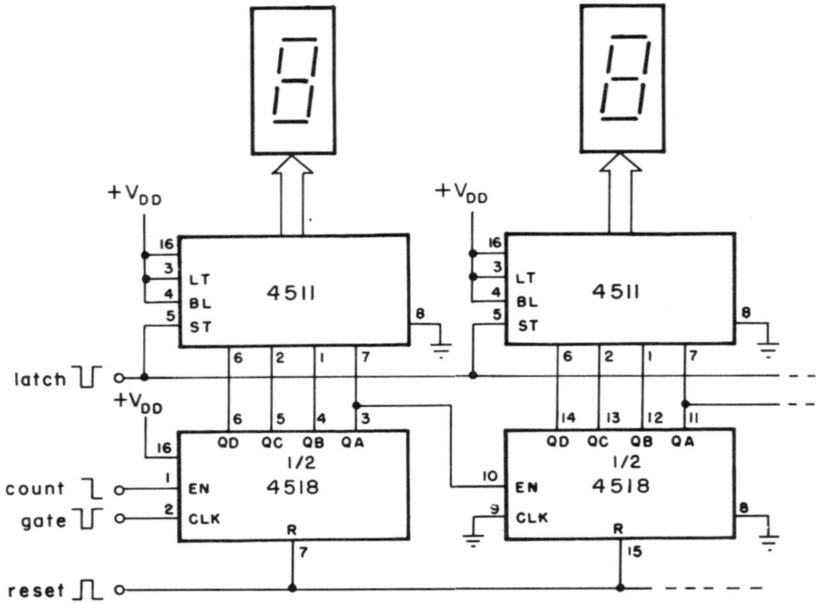

Fig. 6-13. BCD-to-7-segment decoder with latch.

Programmable Up/Down Counters

Up/down counters are often called *add/subtract* counters, in that they are able to add to, as well as subtract from the stored count. This type of counter is particularly useful in applications involving digital frequency synthesizers and the synthesis of arithmetic functions with rate multipliers. In this group of up/down counters are the 4510, 4516, 40192 (74C192), and the 40193 (74C193).

The 4510 (decade) and 4516 (binary or hexadecimal) counters, whose pin diagram is shown in Fig. 6-14, are both 4-bit counters

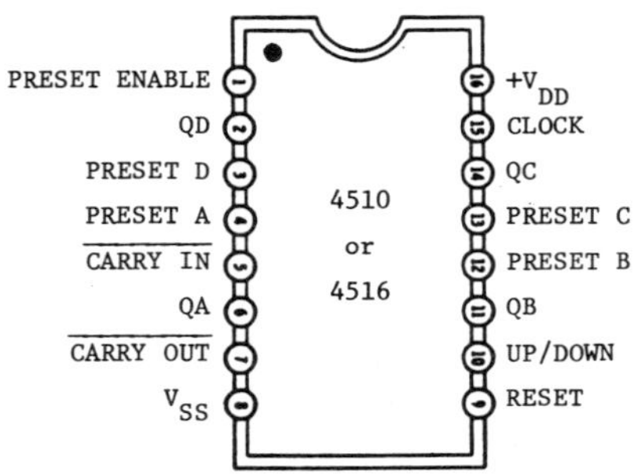

Fig. 6-14. Pin diagram of the 4510 (decimal)/4516 (hexadecimal) programmable up/down counter.

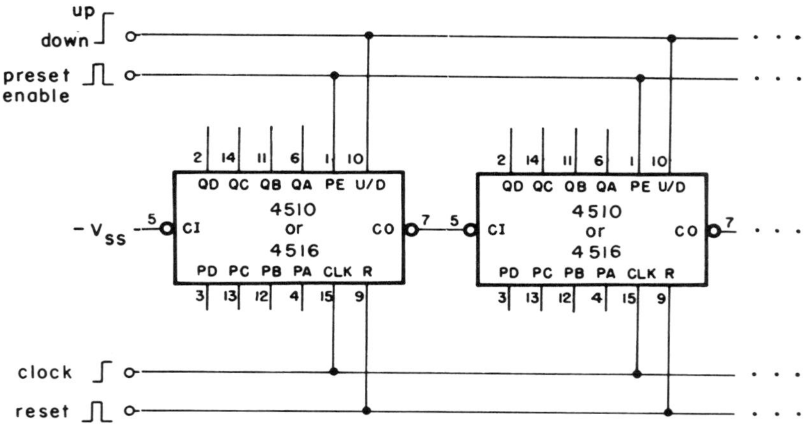

Fig. 6-15. Cascading for parallel clocking.

whose modules can be programmed, or *preset,* to a given value using the standard 4-bit BCD or binary code at the PRESET inputs. Alternative names used for these inputs are *jam* and *load.* The state of the UP/DOWN input determines whether the counter counts up (adds to the count) or down (subtracts from the count) on each positive edge clock transition. If the UP/DOWN input is at logic 1, the counter adds; for a logic 0, the counter subtracts. Table 6-4 gives the truth table for the 4510 and 4516 counters.

As shown in Fig. 6-15, decade or binary counter stages are cascaded for *parallel* clocking by connecting the $\overline{\text{CARRY OUT}}$ signal of the first stage to the $\overline{\text{CARRY IN}}$ pin of the next stage, while the input clock signal simultaneously drives the CLOCK input of all the stages. For *ripple* clocking, the CLOCK and $\overline{\text{CARRY IN}}$ pins of each stage are connected together, as shown in Fig. 6-16.

The 4-bit code is *parallel-loaded* into the counter by momentarily bringing the PRESET ENABLE input at logic 1. If the *units* counter of Fig. 6-16 is preset with the 4-bit code $DCBA_1 = 0100$ (decimal

Table 6-4. 4510/4516 Truth Table

CLOCK	CARRY IN	UP/DOWN	PRESET ENABLE	RESET	Result
0 or 1	1	0 or 1	0	0	no count
↑	0	1	0	0	count up
↑	0	0	0	0	count down
0 or 1	0 or 1	0 or 1	1	0	preset with 4-bit code
0 or 1	0 or 1	0 or 1	0 or 1	1	reset

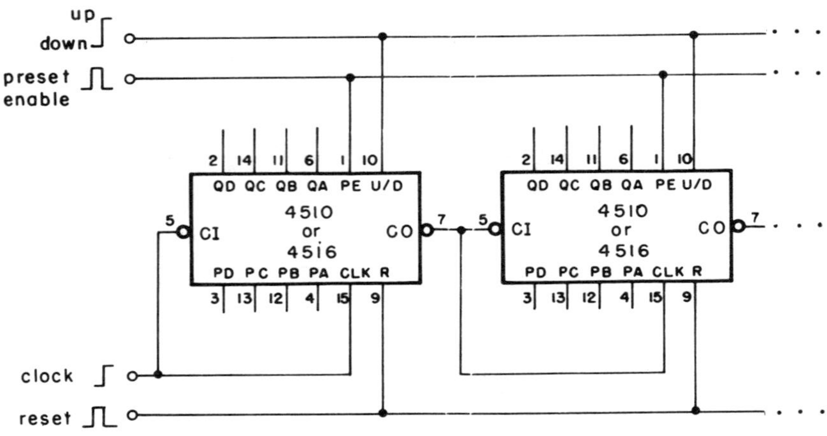

Fig. 6-16. Cascading for ripple clocking.

4), the *tens* counter with $DCBA_{10} = 1000$ (decimal 8), and the *hundreds* counter with $DCBA_{100} = 0011$ (decimal 3), when parallel-loaded, the initial count is then *384*, from which we can count either up or down.

The 4029 counter (Fig. 6-17) is similar in operation to the 4510 and 4516 except that it can be selected for either decade or binary (hexadecimal) counting. A logic 1 on the BINARY/DECADE input results in 4-bit binary counting. In addition there is no reset input. Consequently, the 4029 can only be reset by presetting the 4-bit PRESET inputs to DCBA = 0000 and momentarily bringing the PRESET ENABLE input at logic 1.

In contrast to the 4029, 4510, and 4516 counters, the 40192, or 74C192 (decade counters), and the 40193, or 74C193 (hexadecimal), programmable up/down counters have two separate clock inputs:

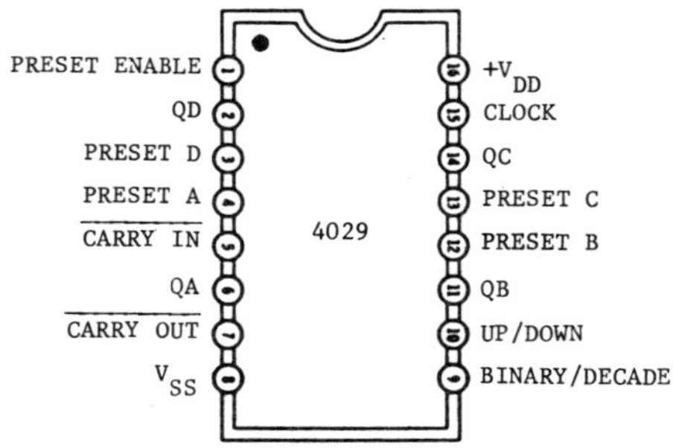

Fig. 6-17. Pin diagram of the 4029 decimal/hexidecimal programmable up/down counter.

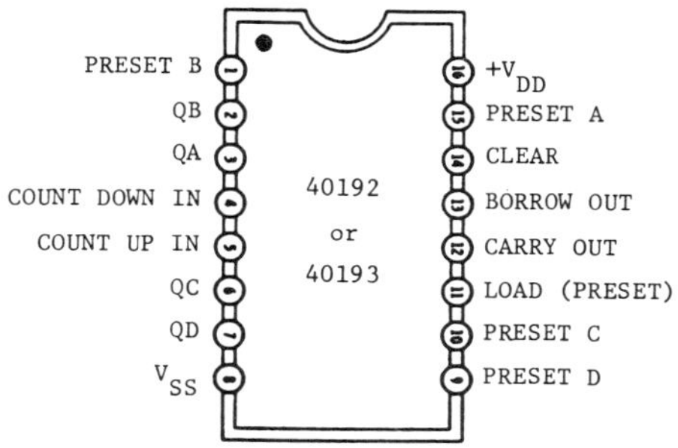

Fig. 6-18. Pin diagram of the 40192 (decimal)/40193 (hexadecimal) programmable up/down counters.

one for counting up and one for counting down, as shown by the common pin diagram of Fig. 6-18.

As illustrated by the two-decade circuit of Fig. 6-19, both the 40192 and 40193 count either up or down on the positive clock edge. When counting, the $\overline{\text{CARRY}}$ and $\overline{\text{BORROW}}$ signals are normally at logic 1. The $\overline{\text{CARRY}}$ signal goes low one-half cycle after the individual counter stage reaches its maximum count when counting *upwards*. The $\overline{\text{BORROW}}$ signal goes low one-half cycle after the individual stage reaches its minimum count when counting *downwards*.

Programmable Down Counters

The 40102 is a 2-decade BCD programmable synchronous down counter in that it internally contains two cascaded 4-bit BCD counters. Depending on the PRESET input settings, the 40102 is able to count down from a maximum count of 99 to 0. The 40103 is a single 8-bit binary counter, which is able to count from a maximum

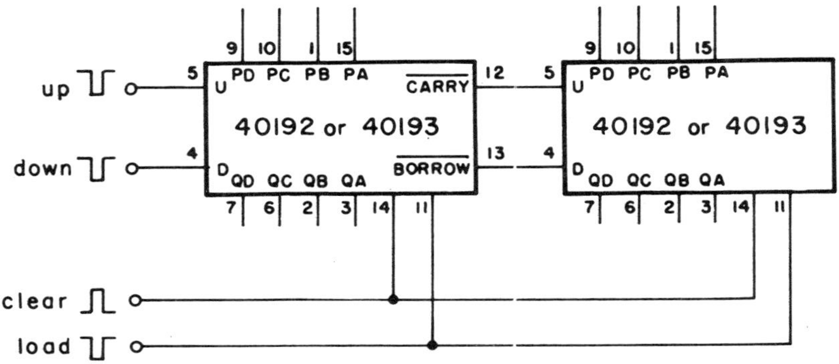

Fig. 6-19. Two-decade up/down counter.

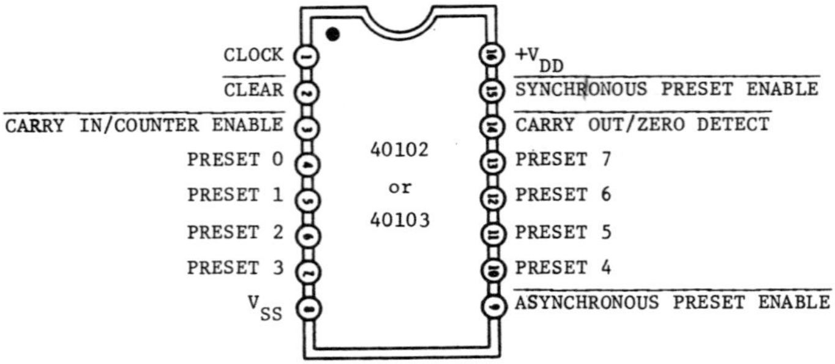

Fig. 6-20. Pin diagram of the 40102/40103 programmable down counters.

count of 255 down to 0. Fig. 6-20 shows the pin diagram common to the 40102 and 40103.

In normal operation, either the 40102 or 40103 is decremented by one count on each positive input clock transition. Down counting is inhibited when the CARRY IN/COUNTER ENABLE (CI/CE) input at logic 1. When the count reaches zero, the CARRY OUT/ZERO DETECT (CO/ZD) output goes low and remains low for one full clock cycle if the CI/CE is also low.

When the SYNCHRONOUS PRESET ENABLE (SPE) input is low, the counter is then parallel-loaded with the settings of the seven PRESET inputs on the *next* positive clock transition, regardless of the state of the CI/CE input. If the ASYNCHRONOUS PRESET ENABLE (APE) input is low, the counter is parallel-loaded asynchronously, regardless of the states of the CLOCK, SPE, CI/CE inputs. When the CLEAR (CLR) input is at logic 0, the 40102 counter is asynchronously cleared to its maximum count of 99, while the 40103 counter is cleared to 255, regardless of the states of the other inputs. All of these actions are summarized in the timing diagram of Fig. 6-21 and listed in Table 6-5.

Fig. 6-22 illustrates how either the 40102 or 40103 is connected as a divide-by-*N* counter. Depending on the settings of the seven PRESET inputs, resulting in a decimal number K, the output frequency will be a factor of 1/(K+1) times the input clock frequency. For the 40102 counter, PRESET inputs 0 through 3 determine the *units* decade, while inputs 4 through 7 determine the *tens* decade. Inputs 3 and 7 are the most significant bits. For the 40103 counter, the count is determined by the settings of an 8-bit binary number. As an example, if the PRESET inputs are set at 01110011, the 40102 will have an output frequency that is 1/74 times the input, while the 40103 will have an output that is 1/116 times the input.

The 4059 programmable divide-by-*N* counter, as shown in Fig. 6-23, can be set to divide an input frequency by any integer between

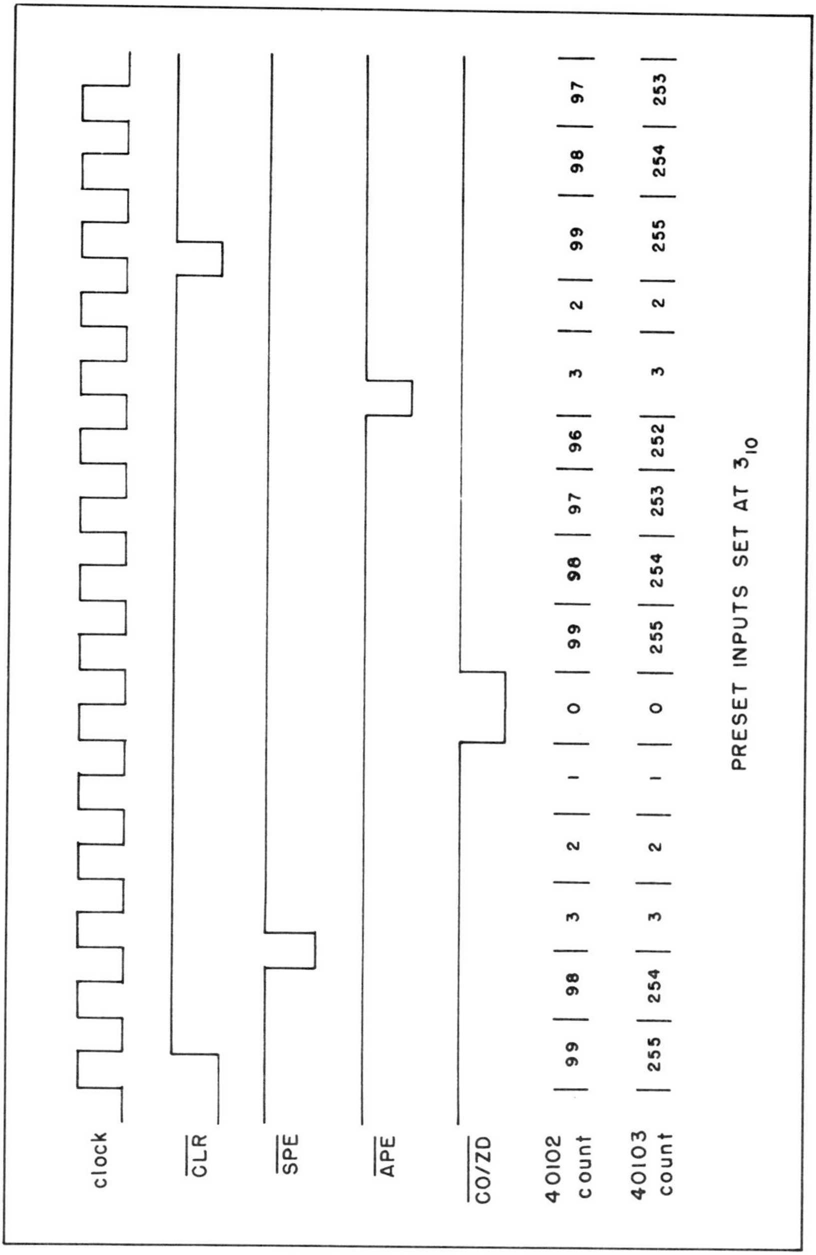

Fig. 6-21. 40102/40103 timing diagram.

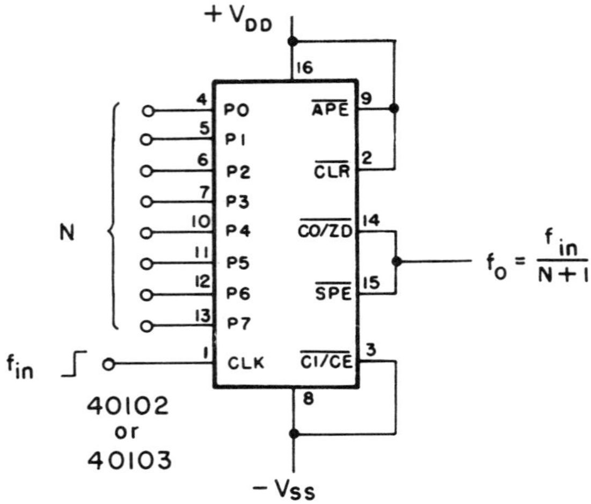

Fig. 6-22. 40102/40103 divide-by-N counter.

Table 6-5. 40102/40103 Truth Table

Control Inputs					
$\overline{CLR}$	$\overline{APE}$	$\overline{SPE}$	$\overline{CI/CE}$	Preset Mode	Result
1	1	1	1	Synchronous	inhibits counting
1	1	1	0	Synchronous	counts down
1	1	0	0 or 1	Synchronous	preset on next positive clock transition
1	0	0 or 1	0 or 1	Asynchronous	preset asynchronously
0	0 or 1	0 or 1	0 or 1	Asynchronous	clear to max count 99 for 40102; 255 for 40103

3 and 15,999. However, the 4059 is not as easily programmed as other types previously discussed. For any setting, the output signal is a pulse that is one input clock cycle wide.

Programming is accomplished with the MODE-SELECT inputs (K_a, K_b, and K_c) which determine the modulus of the first and last counting sections, as illustrated by the block diagram of Fig. 6-24, and summarized in Table 6-6. The MODE-SELECT inputs set the maximum modulus at 9999 when the first counting section divides by either 5 or 10, or set to a maximum of 15,999 when the first section divides by 8, 4, or 2. In addition, the intermediate counting sections can each be preset to a maximum count of 15 instead of 9. A logic 1 at the LATCH ENABLE input causes the output of the

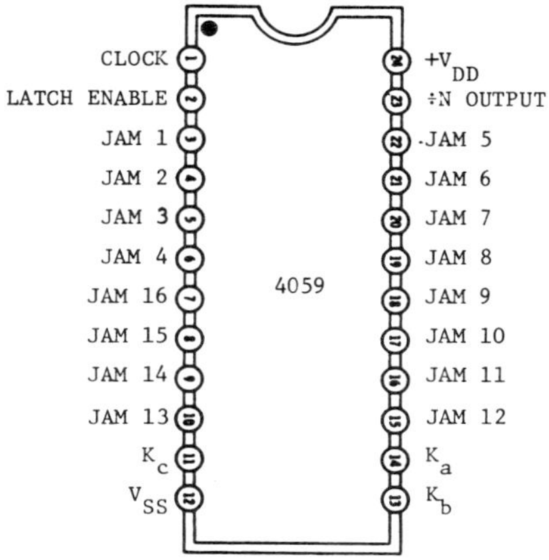

Fig. 6-23. Pin diagram of the 4059 programmable divide-by-N counter.

4059 to remain at logic 1 once an output pulse occurs, and remains at logic 1 until the LATCH ENABLE input returns to logic 0.

To program the 16 PRESET or JAM, and the 3 MODE SELECT inputs, we shall consider the following two examples.

The modulus is to be 6543. Suppose that the MODE of the first counting section is 5, so that 6543 ÷ 5 gives 1308 with a remainder of 3. Then from Table 6-6 with a MODE SELECT of 5, we have:

MODE SELECT (5)	*JAM INPUTS*	
$K_a = 1$	J1 = 1	Remainder (3)
$K_b = 0$	J2 = 1	
$K_c = 1$	J3 = 0	
	J4 = 1	MSB (1)

so that all the JAM inputs are then set at

	Remainder			*MSB*	*Hundreds*				*Tens*				*Units*			
	J3	*J2*	*J1*	*J4*	*J16*	*J15*	*J14*	*J13*	*J12*	*J11*	*J10*	*J9*	*J8*	*J7*	*J6*	*J5*
	0	1	1	1	0	0	1	1	0	0	0	0	1	0	0	0
decimal value		3		1		3				0				8		

The above settings can then be verified as follows:

$$K = [\text{MODE}] \times [(1000 \times \text{MSB}) + (100 \times \text{Hundreds}) + (10 \times \text{Tens}) + (1 \times \text{Units})] + (\text{Remainder})$$

$$= [5][(1000 \times 1) + (100 \times 3) + (10 \times 0) + (1 \times 8)] + (3)$$

$$= 6543$$

Table 6-6. 4059 Counter Truth Table

Mode Select Inputs			1st Counting Section			Last Counting Section			Maximum Modulus
K_a	K_b	K_c	Mode (divides by)	Can be preset to a max of	Jam inputs used	Mode (divides by)	Can be preset to a max of	Jam inputs used	
1	1	1	2	1	1	8	7	2, 3, 4	15,999
0	1	1	4	3	1, 2	4	3	3, 4	15,999
1	0	1	5	4	1, 2, 3	2	1	4	9,999
0	0	1	8	7	1, 2, 3	2	1	4	15,999
1	1	0	10	9	1, 2, 3, 4	1	0	—	9,999
1 or 0	0	0	Master Preset			Master Preset			

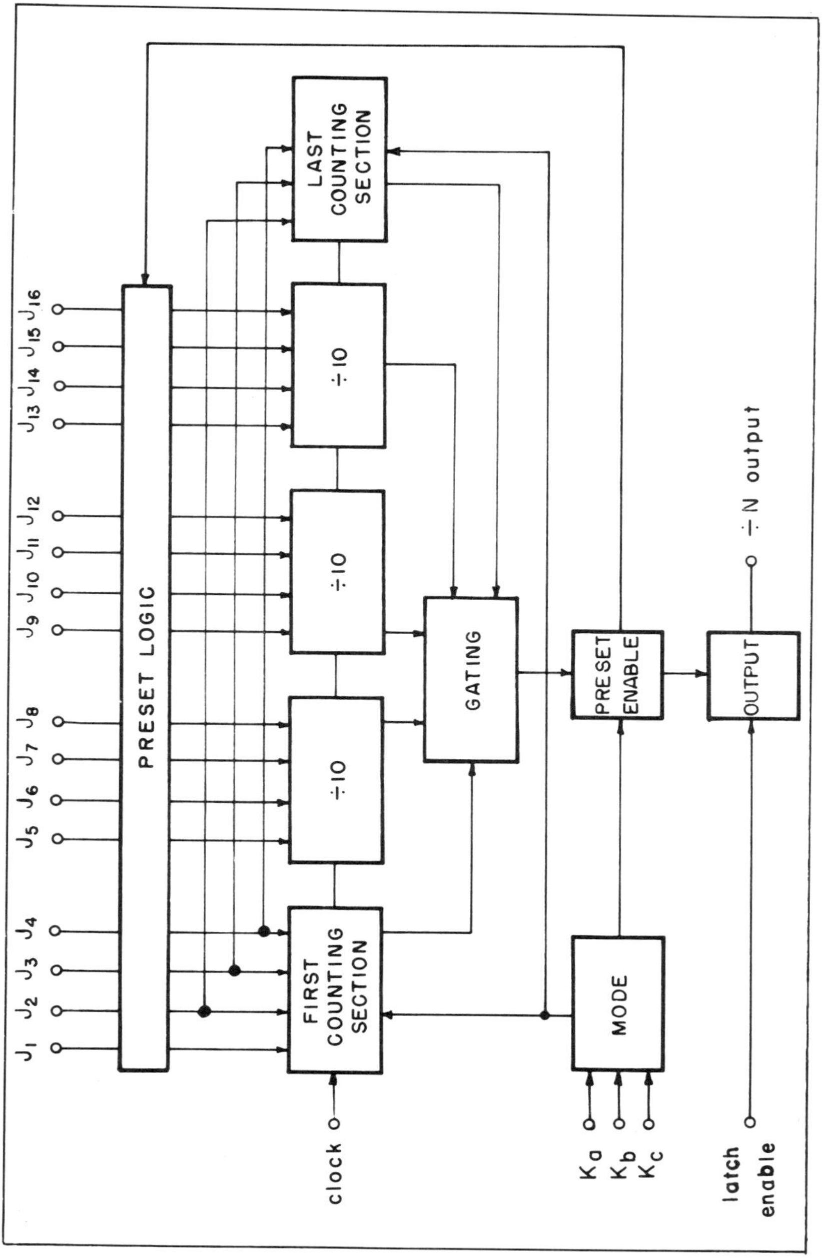

Fig. 6-24. 4059 block diagram.

Consequently, the output frequency will be 1/(6543+1) or 1/6544 times the input. On the other hand, the MODE of the first counter could as easily have been set at 4, so that 6543 ÷ 4 gives 1635 with a remainder of 3. Therefore,

MODE SELECT (4)	*JAM INPUTS*	
$K_a = 0$	$J1 = 1$	Remainder (3)
$K_b = 1$	$J2 = 1$	
$K_c = 1$	$J3 = 0$	
	$J4 = 1$	MSB (1)

so that all the JAM inputs are then set at

	Remainder			*MSB*	*Hundreds*				*Tens*				*Units*			
	J3	*J2*	*J1*	*J4*	*J16*	*J15*	*J14*	*J13*	*J12*	*J11*	*J10*	*J9*	*J8*	*J7*	*J6*	*J5*
	0	1	1	1	0	1	1	0	0	0	1	1	0	1	0	1
decimal value	3			1	6				3				5			

HCTR 0320 Digital Frequency Synthesizer

The HCTR 0320 (Hughes) is a CMOS device which contains both a programmable divider and a phase/frequency detector in a single 28-pin package, as shown in Fig. 6-25. Although the frequency/phase detector is not pertinent to this chapter, a complete discussion on phase detectors and synthesizers is given in *Design*

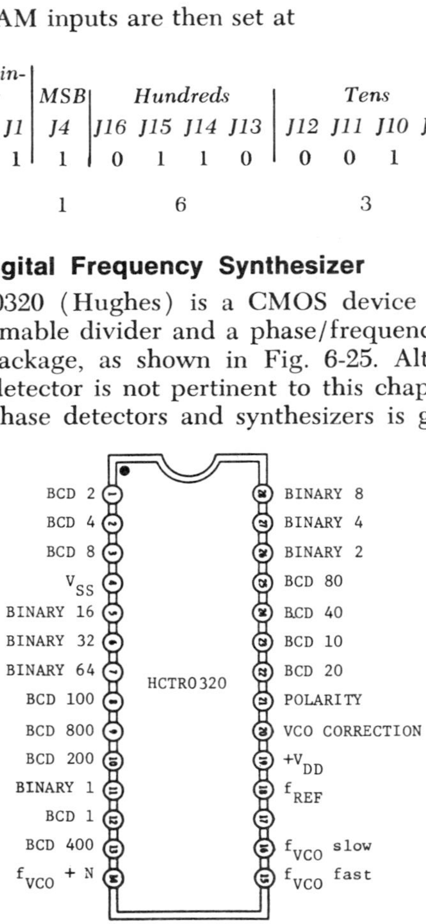

Fig. 6-25. Pin diagram of the HCTR0320 frequency synthesizer.

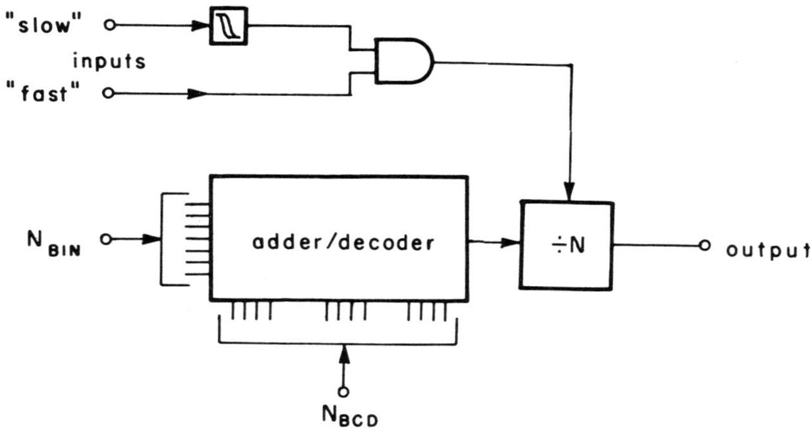

Fig. 6-26. HCTR0320 block diagram.

of Phase-Locked Loop Circuits, with Experiments, published by Howard W. Sams & Co., Inc.

This device, when used as a programmable divide-by-*N* counter, is programmed by adding a three-digit BCD number N_{BCD}, ranging from 0 to 999, to a 7-bit number N_{BIN}, ranging from 0 to 127 (i.e., 0000000 to 1111111). The total modulus N is the sum of N_{BCD} and N_{BIN}, which ranges from 3 to 1023.

As shown in the diagram of Fig. 6-26, the BCD and binary programming inputs are added and decoded, which then feeds a programmable divider. The output of the divider has a frequency that is $1/(N_{BCD} + N_{BIN})$ times the input, and a duty cycle that is 1/N. In addition, there are two types of frequency inputs. The "fast" input at pin 15 is the only TTL-compatible input, and should be used for inputs having fast rise and fall times, or when maximum speed is required. Otherwise, for inputs with slow rise and fall times (e.g., a sine wave), the "slow" input (pin 16) uses an internal Schmitt trigger for proper conditioning. The characteristics of these two inputs are summarized in Table 6-7.

RATE MULTIPLIERS

The *rate multiplier* is a specialized counter circuit that produces an output pulse train whose frequency is proportional to the product of two inputs. Basically, the rate multiplier consists of four JK flip-flops, such as 4027s, a pair of 2-input AND gates (4081), four 4-input AND gates (4082), and a single 4-input OR gate (4072) as shown in Fig. 6-27.

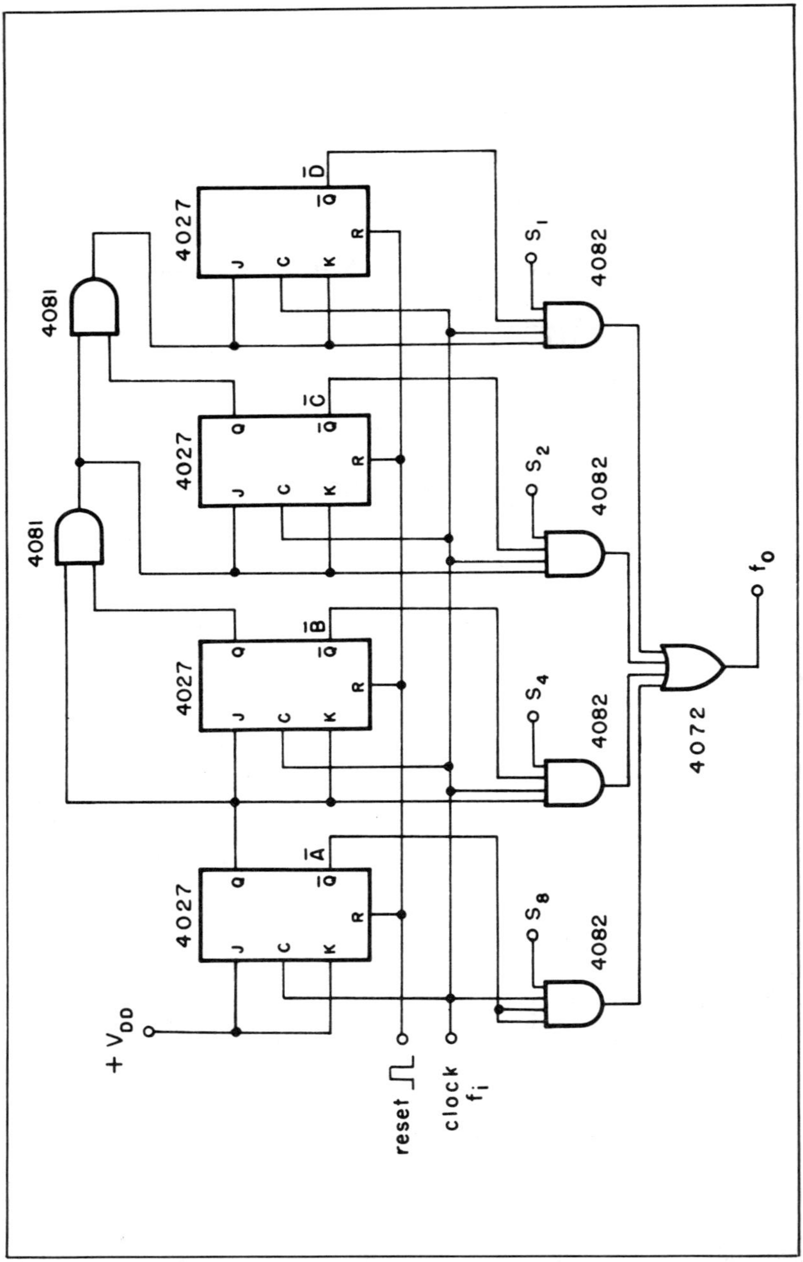

Fig. 6-27. Rate multiplier using logic elements.

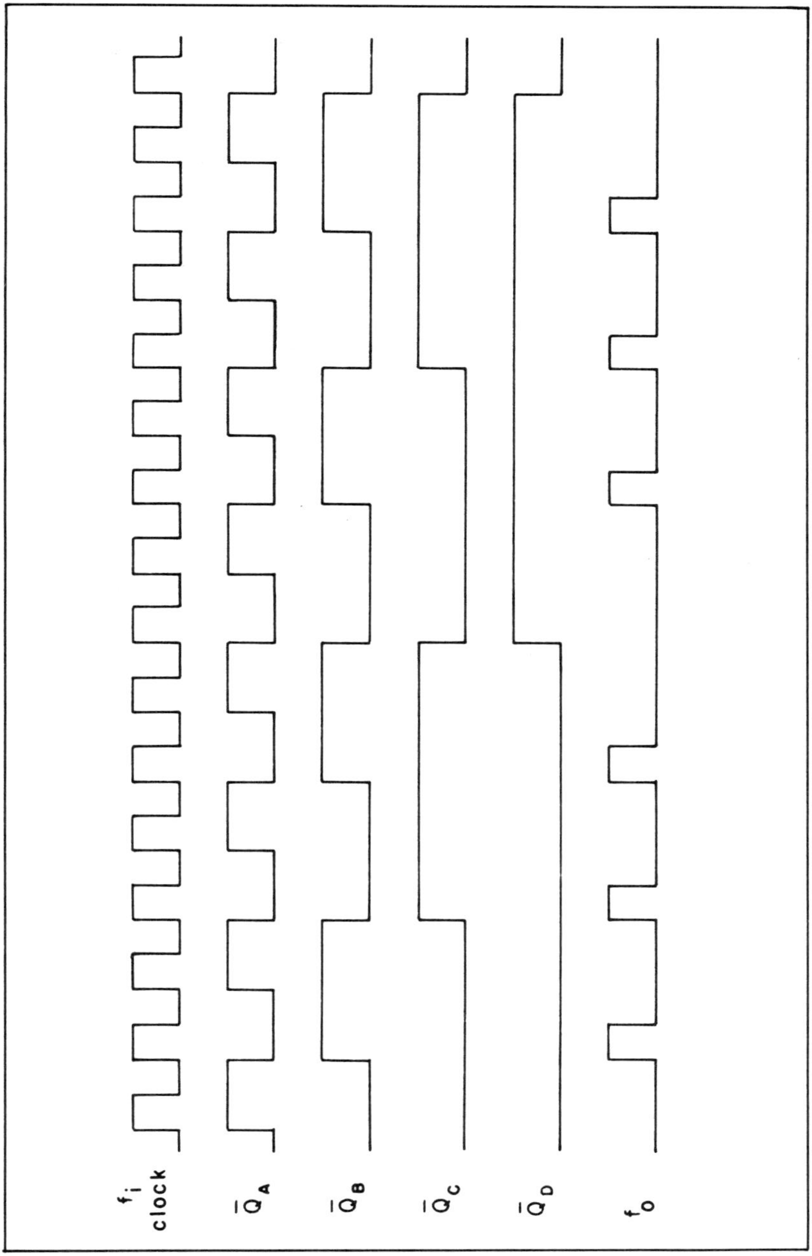

Fig. 6-28. Rate multiplier timing diagram.

One of the two required inputs is a clock frequency f_i, while the other is a preprogrammed 4-bit multiplier, which can be either binary or BCD. For a *binary* rate multiplier, the output frequency f_o will be N/16 times f_i, where N is a 4-bit binary number ranging from 1 to 15. The *BCD* rate multiplier has an output frequency that ranges from 1/10 to 9/10 times the input frequency. In either case, the output rate is always less than the input.

Table 6-7. Input Characteristics of the HCTR 0320

Input	Maximum Frequency		10%–90% Rise and Fall Time	
	$V_{DD} = 5$ V	$V_{DD} = 10$ V	$V_{DD} = 5$ V	$V_{DD} = 10$ V
"Fast" (pin 15)	5 MHz	10 MHz	100 ns	50 ns
"Slow" (pin 16)	2.5 MHz	5 MHz	No Limit	No Limit

The timing diagram in Fig. 6-28 shows how a value for $f_o = (6/16)f_i$ is obtained when the 2^2 and 2^1 programming inputs are at logic 1 and the 2^3 and 2^0 inputs are logic 0. From the timing diagram, the output of the rate multiplier consists of pulses that are *unevenly spaced,* although there are 6 output pulses for every 16 input cycles. Consequently, it must be recognized that the output rate is an *average* rate, so that there is always a *round-off* error.

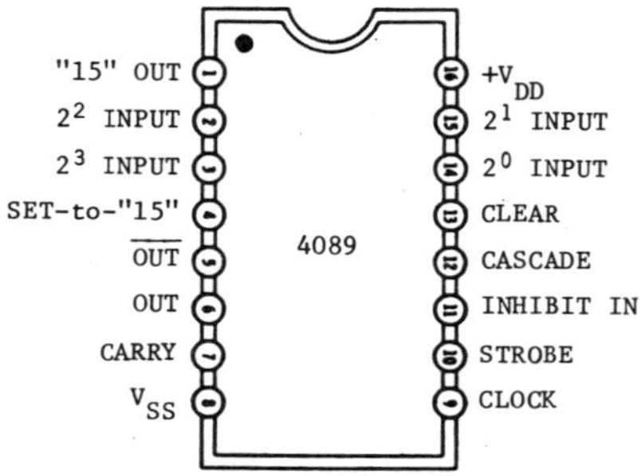

Fig. 6-29. Pin diagram of the 4089 binary rate multiplier.

However, this error can be reduced by increasing the bit capacity of the multiplier.

The 4089 (Fig. 6-29) is a binary rate multiplier, while the 4527 (Fig. 6-30) is a BCD multiplier. Each contains an internal synchronous 4-bit counter which, together with the settings of the four binary or BCD RATE SELECT inputs, produces a specified output pulse train. The INHIBIT IN, CASCADE, STROBE, CLEAR, and

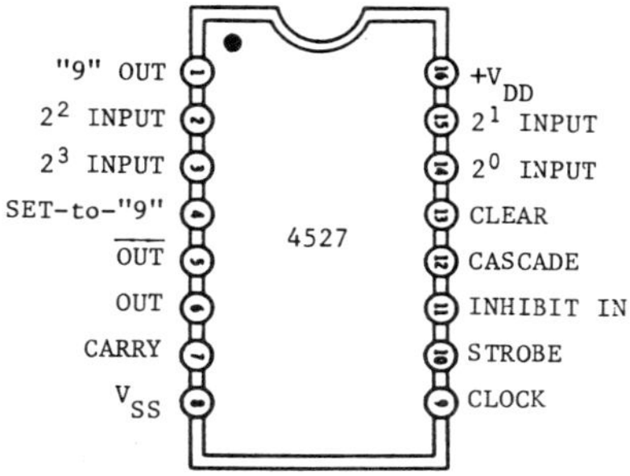

Fig. 6-30. Pin diagram of the 4527 BCD rate multiplier.

SET-to-"9" (for the 4527) or SET-to-"15" (for the 4089) inputs are normally at logic 0.

As shown in Fig. 6-31, two or more 4089 binary rate multipliers are connected in *series* for which the output rate is given by the generalized relation

$$f_o = \frac{A \cdot B \cdot C \cdot \cdots \cdot N}{2^n}(f_i) \qquad \text{(Eq. 6-1)}$$

The numbers A, B, etc., are the decimal equivalents of the 4-bit code for each 4089 section, and n is the number of bits. If three 4089

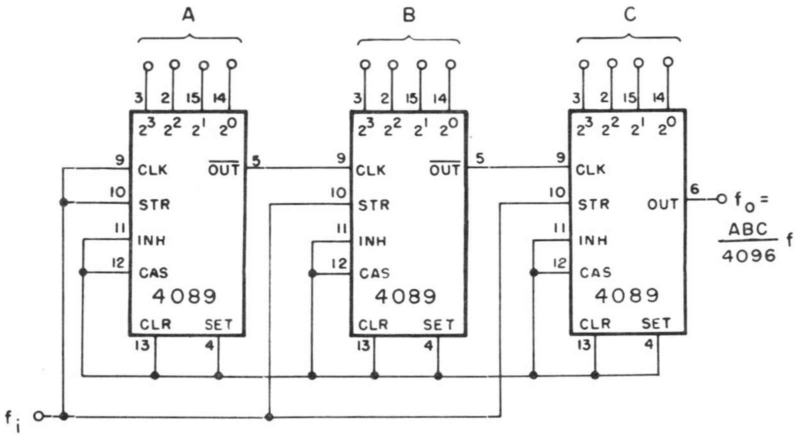

Fig. 6-31. Rate multipliers connected in series.

devices are connected in series, then n is equal to $2^{[(3)(4)]}$ or 4096. Then the numbers A, B, and C individually would be any number between 1 and 15. Consequently, the maximum fraction of the input frequency in this case is 3375/4096.

Using the 4527, the devices are connected in *parallel,* as shown in Fig. 6-32 forming a multibit multiplier. The output rate is

$$f_o = \frac{A}{10^n}(f_i) \qquad \text{(Eq. 6-2)}$$

where A is the BCD number and *n* is the number of parallel stages. If two 4527 rate multipliers are connected in parallel, then A can range from 1 to 99, and the denominator of Equation 6-2 equals 100. Consequently, the output rate can vary from 0.01 to 0.99 times the input rate f_i, depending on the settings of the programming inputs. The following circuits will illustrate how rate multipliers can be used to perform a variety of mathematical functions. Additional details concerning the 4089 and 4527 rate multipliers are fully de-

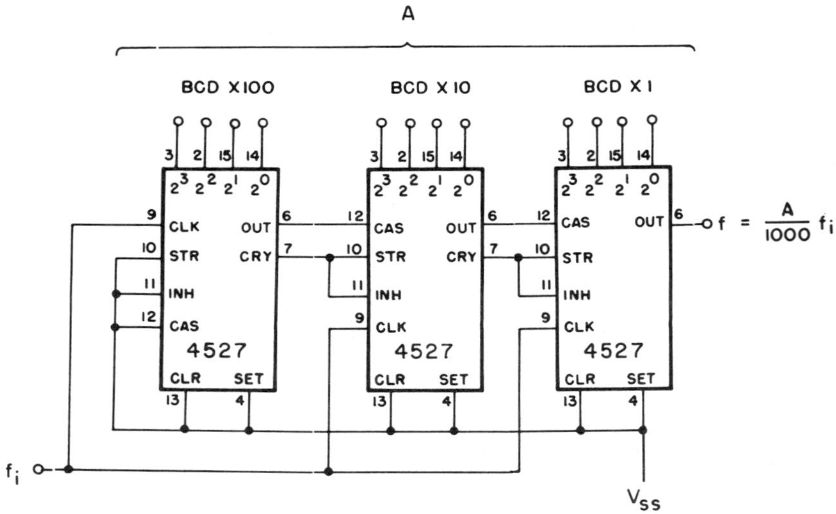

Fig. 6-32. Rate multipliers connected in parallel.

scribed in the RCA Application Note ICAN-6739, *COS/MOS Rate Multipliers-Versatile Circuits for Synthesizing Digital Functions,* by Gary J. Summers.

Multiplication

By using three or more rate multipliers and an up/down counter (4510 or 4516), we are able to obtain the product of two or more variables. The product is not in terms of an output frequency, but rather a binary, or BCD, number which can be decoded to drive a 7-segment LED display.

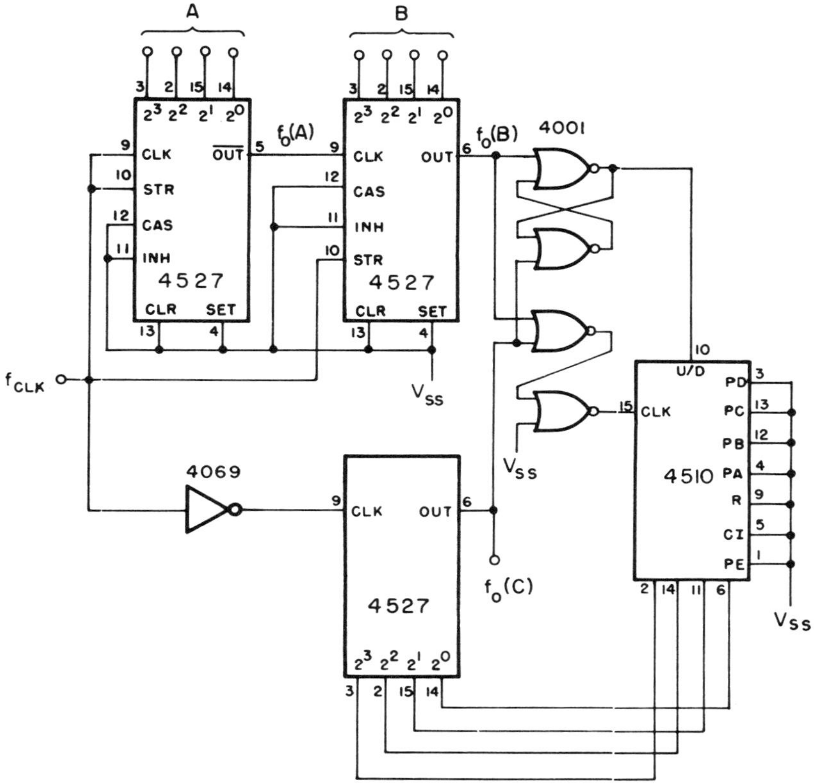

Fig. 6-33. Multiplication of two 4-bit numbers.

With the circuit of Fig. 6-33, we can program two 4-bit variables, A and B (binary or BCD), so that the 4-bit output of the up/down counter is

$$N = \frac{A \cdot B}{k} \qquad \text{(Eq. 6-3)}$$

where k is 10 when using the 4527, and equal to 16 for the 4089 rate multiplier. The analysis of the operation of this circuit can be understood, assuming that 4527 BCD rate multipliers and the 4510 BCD up/down counter are used. The output of the 4527 rate multiplier programmed with the 4-bit BCD number "A" is

$$f_o(A) = \frac{A}{10}(f_{CLK}) \qquad \text{(Eq. 6-4)}$$

Similarly, the output of the multiplier programmed with the 4-bit BCD number "B" is

$$f_o(B) = \frac{B}{10} f_o(A)$$

$$= \frac{A \cdot B}{100} (f_{CLK}) \qquad \text{(Eq. 6-5)}$$

It should be noted that these two rate multipliers are connected in series, as was shown previously in Fig. 6-31, but using 4089s. The output rate of the remaining 4527 is

$$f_o(C) = \frac{N}{10} (f_{CLK}) \qquad \text{(Eq. 6-6)}$$

where the number N is determined by the 4-bit output of the 4510 up/down counter (Equation 6-3), or

$$f_o(C) = \frac{A \cdot B}{10} (f_{CLK})$$

The top pair of NOR gates (4001), which are cross-coupled, convert the outputs of the "A" and "C" rate multipliers to a signal that tells the 4510 counter whether to count up or down. The "B" output controls the up count, while the "C" output controls the down count. As these two outputs change, the 4510 counts up or down until the loop stabilizes, in which case

$$f_o(B) = f_o(C)$$

$$\frac{A \cdot B}{100} (f_{CLK}) = \frac{N}{10} (f_{CLK})$$

from which it can be seen that

$$N = \frac{A \cdot B}{10} \qquad \text{(Eq. 6-7)}$$

Frequency Ratios

By using a pair of BCD (or binary) rate multipliers and a BCD (or binary) up/down counter in a closed loop, as shown in Fig. 6-34, we are able to compute the ratio of two input frequencies, f_1 and f_2. Such a technique is the basis for several measurement systems, such as a "miles/gallon" indicator.* Since the two input frequencies are usually asynchronous, care must be taken to make certain that the two inputs do not coincide. Fig. 6-35 shows a simple

* Summers, G. J., "Miles/Gallon Measurement Made Easy with CMOS Rate Multipliers," *EDN*, January 20, 1976, pp. 61–63.

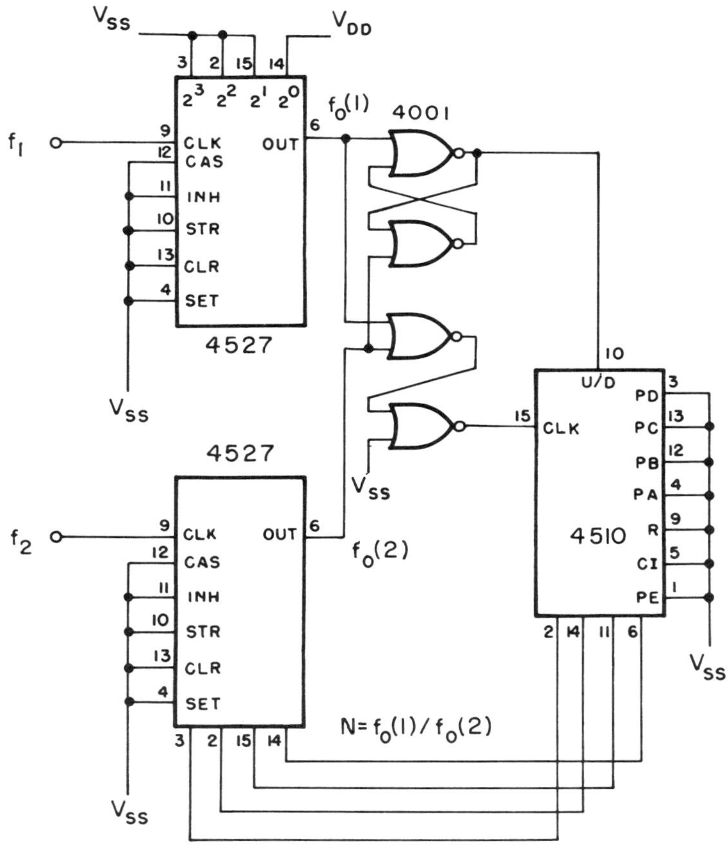

Fig. 6-34. Determining the ratio of two frequencies.

circuit that eliminates this possibility, using a clock frequency that is at least four times faster than the higher of f_1 or f_2.

One of the rate multipliers, whose program inputs are set at DCBA = 0001, or a decimal "1," the output rate is then

$$f_o(1) = \frac{f_1}{10} \qquad \text{(Eq. 6-8)}$$

The output rate of the second rate multiplier, whose program inputs are set by the 4-bit output N of the up/down counter, is

$$f_o(2) = \frac{Nf_2}{10} \qquad \text{(Eq. 6-9)}$$

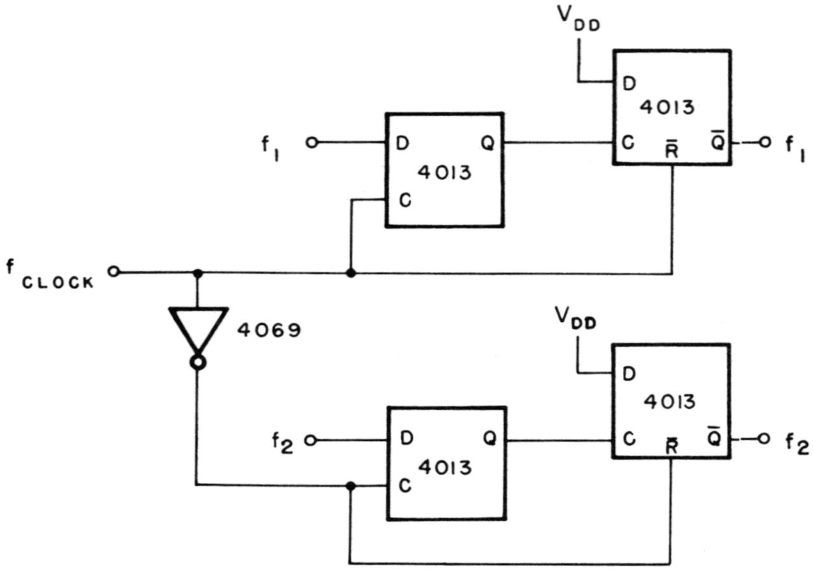

Fig. 6-35. Elimination of waveform coincidence.

When the loop finally stabilizes, the two output rates are equal, so that

$$f_o(1) = f_o(2)$$
$$\frac{f_1}{10} = \frac{Nf_2}{10}$$

or

$$N = \frac{f_1}{f_2} \qquad \text{(Eq. 6-10)}$$

WAVEFORM GENERATION

Although operational amplifiers and other specialized integrated circuits are frequently used to generate periodic waveforms such as sine waves, triangles, and sawtooths, CMOS counters can also be used to generate periodic waveforms with respectable results.

The basis for the digital generation of sine waves was described in a paper by A. C. Davies* using a nonrecursive or transversal digi-

* Davies, A. C., "Digital Generation of Low-Frequency Sine Waves," *IEEE Transactions on Instruments and Measurement,* Vol. IM-18, No. 2, June, 1969, pp. 97–105.

tal filter. With binary ripple and walking-ring (Johnson) type CMOS counters, the same results can be obtained. By properly adding the outputs of the counter, which are some binary fraction of the input with various delays, the principle harmonics of the input square-

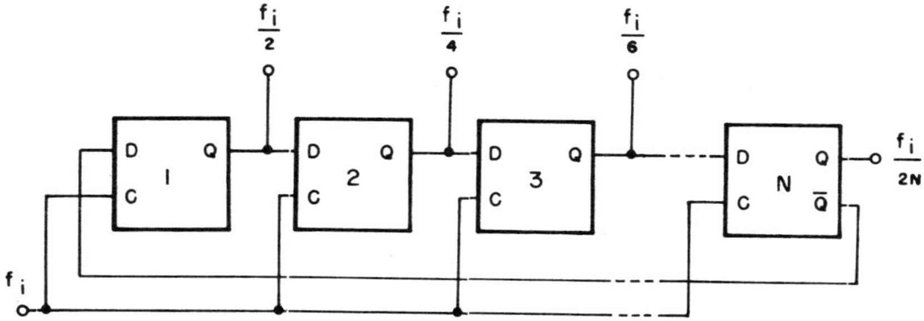

Fig. 6-36. Multistage walking-ring counter.

wave clock can be filtered out, thus leaving only the fundamental component with several higher-order harmonics. The output is then a "stepped" or "chunky" approximation of a sine wave.

As shown in Fig. 6-36, a multistage walking-ring counter can be made from D-type flip-flops, such as the 4013 or 74C74. For N flip-flops (stages), the output frequency at any one of the flip-flop's Q

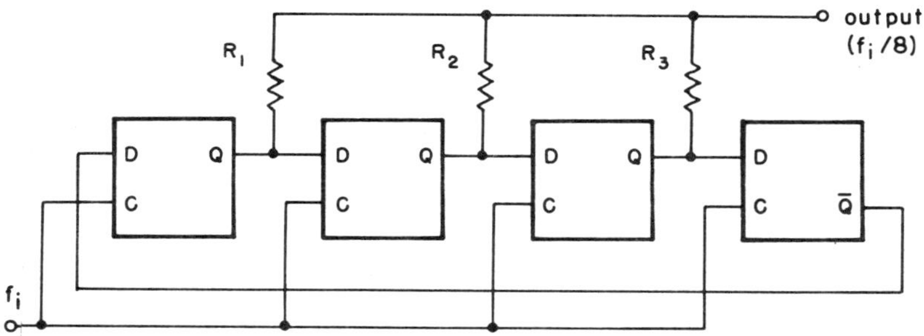

Fig. 6-37. Four-stage sine-wave generator.

outputs will be 1/2N times the input frequency. Consequently, if 5 stages are used, then the output frequency will be 1/10 the input. As the input clock frequency changes, so does the output. Using resistors, the outputs of N-1 stages are then summed to generate the sine wave, as shown for a 4-stage counter in Fig. 6-37. Based on

Davies' mathematics, the weighting resistors are determined by the generalized formula

$$R_N = \left(\frac{\frac{\sin(k)\pi}{N}}{\sin(\pi/N)}\right)(R_1) \qquad k = 1, 2, 3, 4 \ldots (N\text{-}1) \qquad \text{(Eq. 6-11)}$$

The lowest uncancelled harmonic is the $(2N - 1)$th which has an amplitude of

$$dB = 20 \log_{10}\left(\frac{1}{(2N - 1)}\right) \qquad \text{(Eq. 6-12)}$$

relative to the fundamental. For the 4-stage sine-wave generator shown in Fig. 6-37, the values of the resistors relative to R_1 are then

$$R_1/R_1 = 1.000$$
$$R_2/R_1 = 1.414$$
$$R_3/R_1 = 1.000$$

The lowest uncancelled harmonic is the 7th. If the input clock frequency is 10 kHz, then the output sine wave will have a fundamental component of 10 kHz/8 or 1250 Hz, and the lowest uncancelled harmonic is the 7th, which is a frequency of 8750 Hz. At the 7th harmonic, the amplitude will be 1/7th of the fundamental, or −16.9 dB. For most circuits, 5% resistors are adequate. If a 27-kΩ resistor is used for both R_1 and R_3, then a 39-kΩ resistor is required for R_2.

Instead of using discrete D-type flip-flops, walking-ring CMOS counters, such as the 4018, can be used, as shown by the 5-stage circuit of Fig. 6-38A. However, as seen in Fig. 6-38B, the resemblance to a sine wave is not very good. One possible solution is to increase the number of stages, thereby increasing the number of steps for one cycle, but this method has the disadvantage of adding more resistors in addition to more counters or flip-flops. Even still, the resulting waveform will possess some "rough edges."

An alternate approach is to add an active low-pass filter to the output of the generator, as shown in Fig. 6-39. The cutoff frequency of the filter should be well below the lowest uncancelled harmonic so that it will be essentially eliminated from the output sine wave. Information on the design of active filters can be found in either the *Design of Active Filters, with Experiments*, or the *Active Filter Cookbook*, both published by Howard W. Sams & Co., Inc.

Besides using a 4018 counter, binary ripple counters such as the 4024, or any of the others listed in Table 6-1, can be used to generate sine waves. As shown in Fig. 6-40, a 4024 counter, using the first 3 stages, with a pair of 4070 exclusive-OR gates to form a 4-stage generator. Note that only two resistors are required instead of three. An 8-stage circuit using a 4024 counter, 4070 exclusive-OR gates and a 4051 analog multiplexer is shown in Fig. 6-41, which was described in a 1975 issue of *Electronics.**

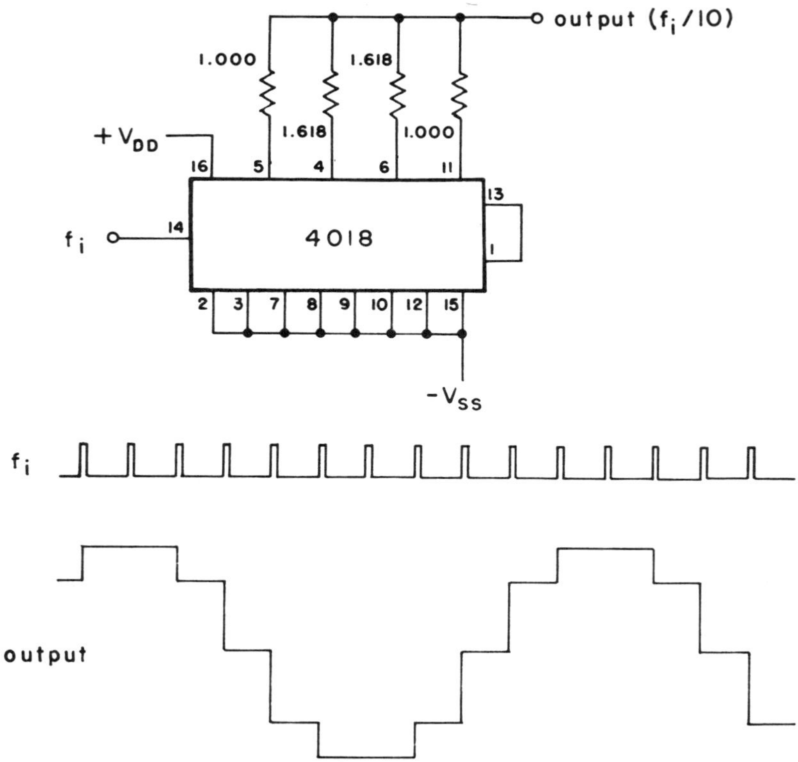

Fig. 6-38. Five-stage sine-wave generator using a 4018 counter.

Using a 4029 counter, as shown in Fig. 6-42, it is possible to generate triangle waveforms, with the help of a R-2R ladder network. As the 4029 is clocked up to its final count, determined by the settings of the PRESET inputs, the CARRY OUT (CO) signal goes from high to low, which is then inverted and used to clock the D-type flip-flop. The clocking action changes the mode of the 4029 to count either up or down. When the count reaches zero (DCBA = 0000), the CO signal clocks the flip-flop, so that the counter counts up every other time the flip-flop is clocked.

---

* McGuire, P. L., "Digital Pulses Synthesize Audio Sine Waves," *Electronics*, October 2, 1975, pp. 104–105.

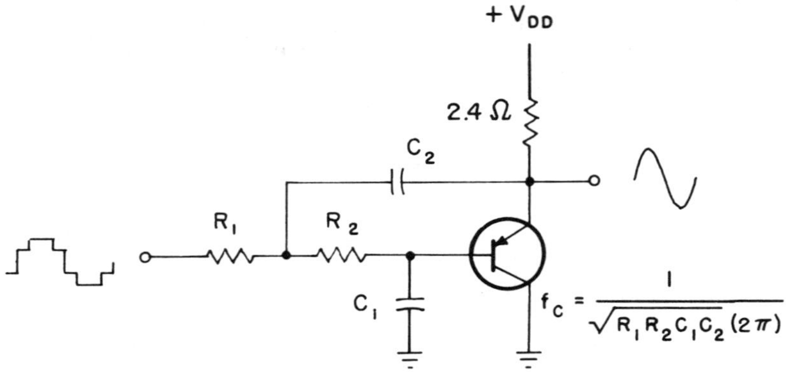

(A) 2nd-order transistor circuit.

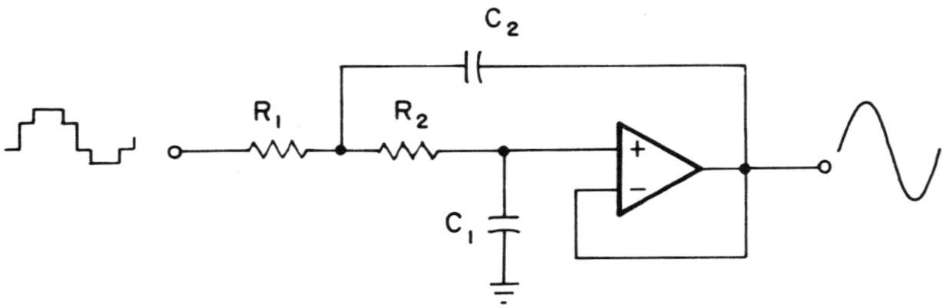

(B) 2nd-order op-amp circuit.

Fig. 6-39. Active low-pass filters.

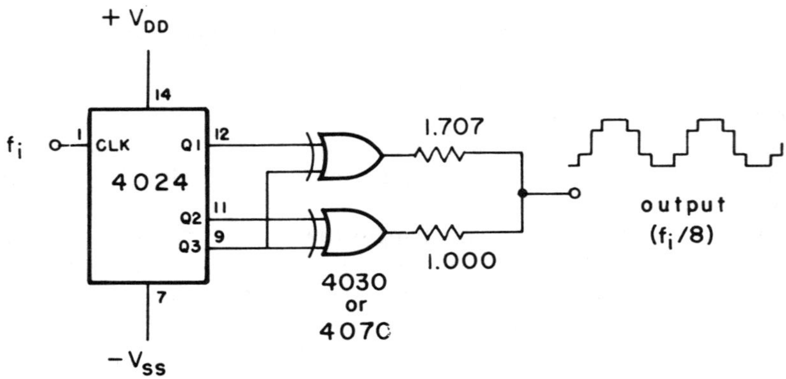

Fig. 6-40. Four-stage sine-wave generator using a 4024 binary ripple counter.

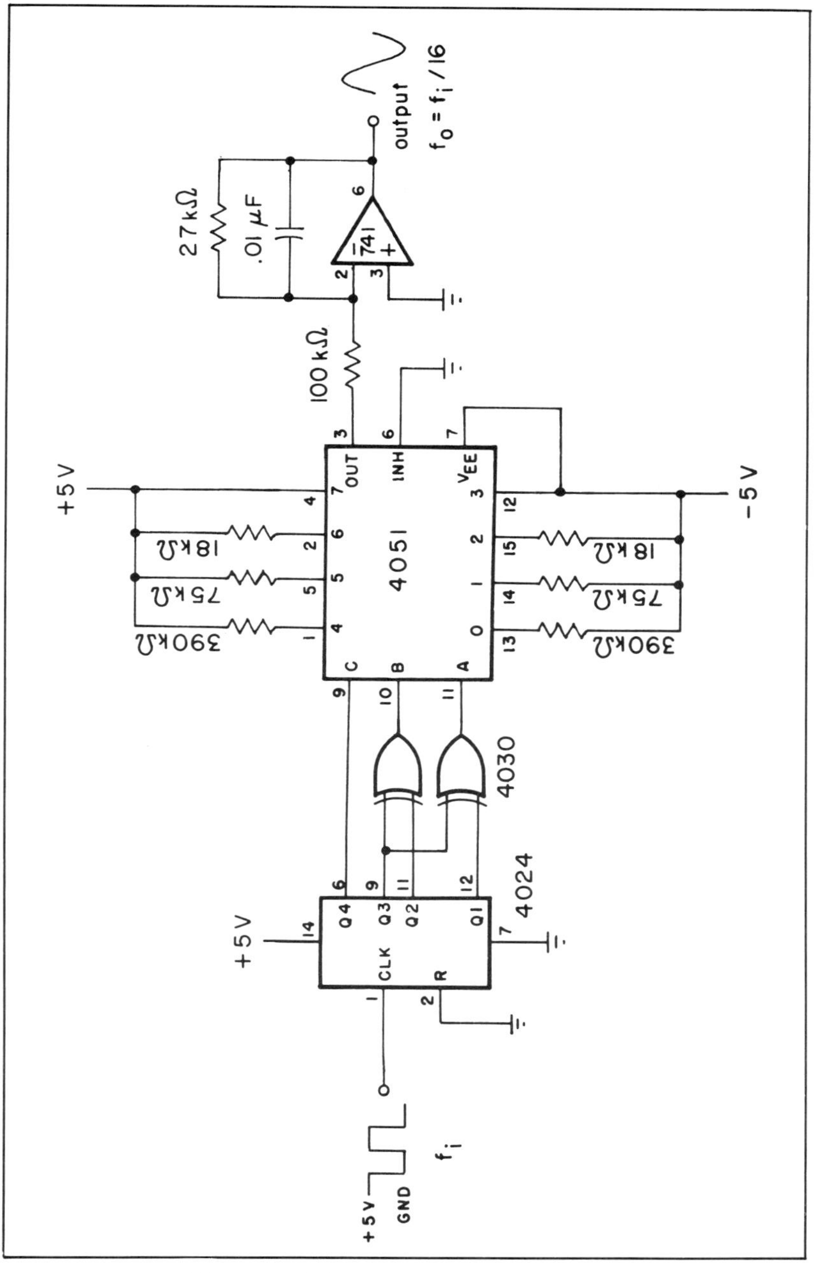

Fig. 6-41. Eight-stage sine-wave generator.

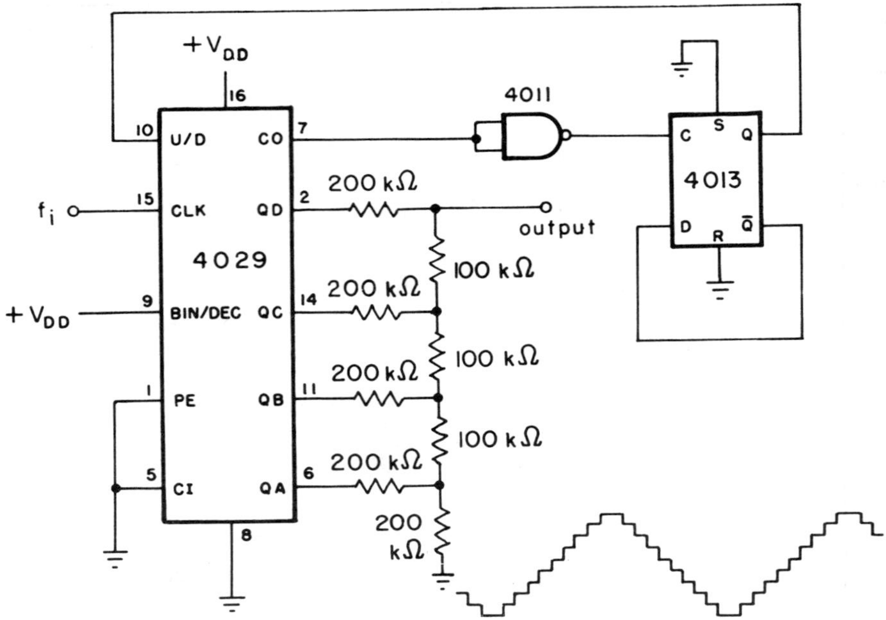

Fig. 6-42. Triangle waveform generator.

As shown in Fig. 6-43, a R-2R ladder network used with 4016 transmission gates, 4049 inverters, and a 4520 binary up-counter to generate a sawtooth waveform in 16 equal voltage steps from ground to the positive supply voltage.

AN INTRODUCTION TO THE EXPERIMENTS

The following experiments are designed to demonstrate the operation of several types of binary ripple and synchronous counters, fixed and programmable. In addition, it will be demonstrated how walking-ring counters, such as the 4018, can be wired to generate periodic waveforms of known frequency, such as sine waves. The experiments that you will perform can be summarized as follows:

Experiment No.	*Purpose*
1	Demonstrates the operation of a 4022 1-of-8 counter/divider.
2	Demonstrates how the 4017 1-of-10 counter can be wired to divide an input frequency by any integer from 2 to 10.

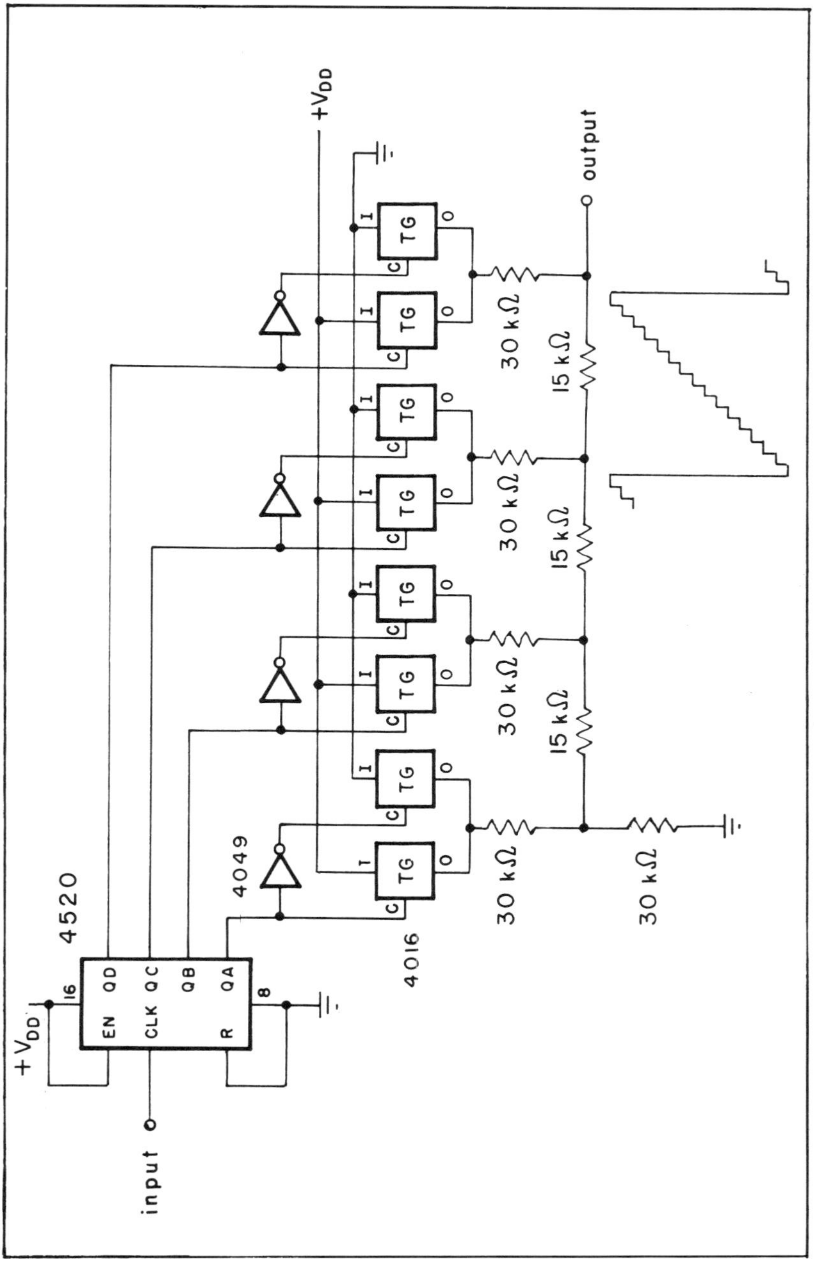

Fig. 6-43. Sawtooth waveform generator.

3	Demonstrates the operation of the 40192/74C192 programmable up/down counter.
4	Demonstrates the operation of the Hughes HCTR0320 frequency synthesizer as a programmable counter.
5	Demonstrates the design and operation of a simple sine-wave generator made from a 4018 5-stage Johnson counter.

EXPERIMENT 1

Purpose

The purpose of this experiment is to demonstrate the operation of the 4022 1-of-8 counter/divider.

Pin Configuration of Integrated Circuit Chip (Fig. 6-44)

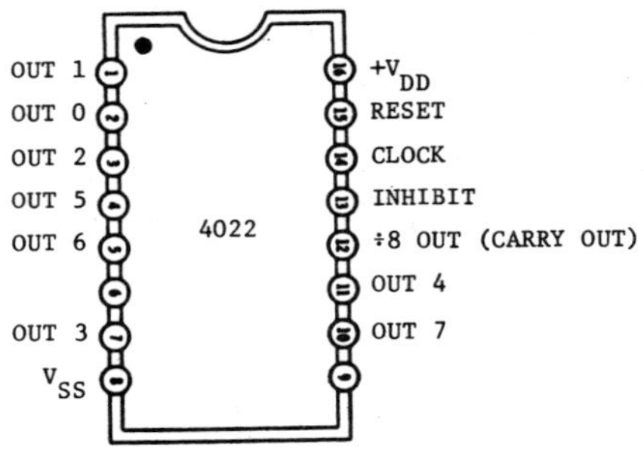

Fig. 6-44.

Schematic Diagram of Circuit (Fig. 6-45)

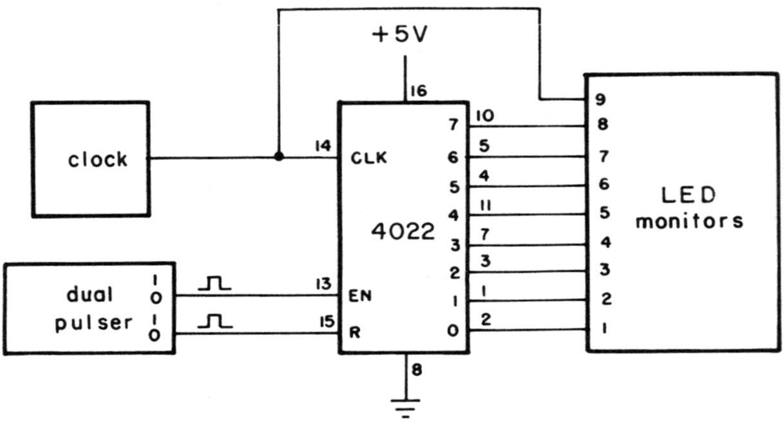

Fig. 6-45.

Step 1

Wire the circuit as shown in the schematic diagram. You will need nine LED monitors, one for each decoded output and the clock input.

Step 2

Apply power to the breadboard and adjust the input clock frequency so that LED monitor No. 9 flashes approximately once each second. Describe what happens to the eight LED monitors connected to the decoded outputs.

The LED monitors sequentially light up for one cycle of the input clock. After LED monitor No. 8 goes out, LED monitor No. 1 lights up and the cycle is repeated. The eight decoded outputs are normally at logic 0, so that the LED monitors are unlit, and the outputs go high only at their respective decoded time slots, and remain high for one full clock cycle.

Step 3

During the counting sequence, press the pulser that is connected to the RESET of the 4022, so that pin 15 is at logic 1. What do you observe on the LED monitors?

When the pulser is pressed, only LED monitor No. 1 should be lit, indicating that the counter has been reset to the zero count, regardless of which LED monitor was lit before the pulser was pressed. Try this several times. You should observe that each time the pulser is released, the count begins at zero, for which LED monitor No. 1 is lit.

Step 4

While the LED monitors are flashing, press the pulser that is connected to the CLOCK ENABLE input, so that pin 13 is at logic 1. What do you observe on the LED monitors?

When you press in the pulser, the LED monitors stop sequencing. Now release the pulser. What happens?

When the pulser is released, the LED monitors continue their orderly sequence from the point that the input of the counter was disabled. Consequently, a logic 1 inhibits, or prevents, the counter from continuing its count sequence.

EXPERIMENT 2

Purpose

The purpose of this experiment is to demonstrate how the 4017 1-of-10 counter can be wired to divide an input frequency by any integer from 2 to 10.

Pin Configuration of Integrated Circuit Chip (Fig. 6-46)

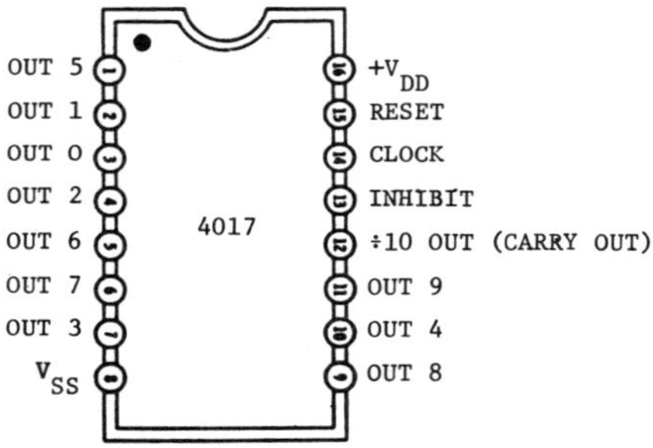

Fig. 6-46.

Schematic Diagram of Circuit (Fig. 6-47)

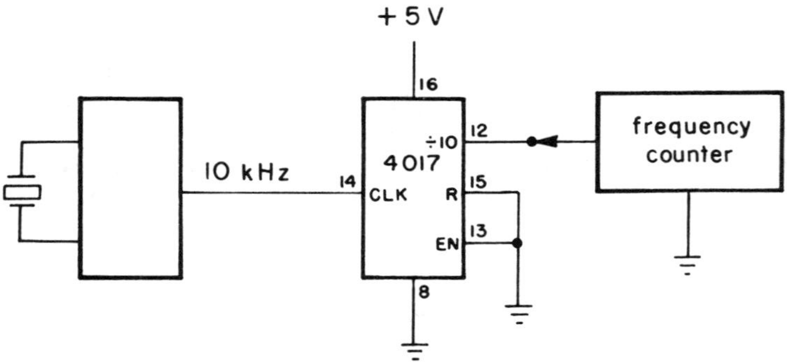

Fig. 6-47.

Step 1

Wire the circuit shown in the schematic diagram and then apply power to the breadboard. If you are not using a 10-kHz crystal controlled frequency reference, use a square wave from a function generator, whose voltage varies between ground and +5 volts, and adjust the frequency with a frequency counter as close as possible to 10 kHz at pin 14. With your frequency counter, what is the output frequency at pin 12?

The output frequency should be 1.0 kHz, or 1/10 the input frequency, since pin 12 is the ÷10 OUTPUT.

Step 2

Now sequentially connect the frequency counter to pins 1, 2, 3, 4, 5, 6, 7, 9, 10, and 11. What do you measure?

You should measure at each pin the same frequency that you measured in Step 2, or 1.0 kHz. The only difference between these outputs and the output at pin 12 is that the output at pin 12 is a symmetrical square wave, while the output of these pins is at logic 1 for the duration of one input cycle, or a 10% duty cycle.

Step 3

Disconnect the frequency reference from the circuit and then disconnect the power from the breadboard. Now connect the OUT 2 pin (pin 4) to the RESET pin (pin 15), and connect the frequency counter to the OUT 0 pin (pin 3). Apply power to the breadboard and reconnect the frequency reference to pin 14 as before. What is the measured frequency at pin 3?

The frequency should be 5.0 kHz, or one-half the input frequency. Starting with a count of zero, each decoded output sequentially goes high for one input cycle, as was shown in the previous experiment using the 4022 1-of-8 counter. After two input cycles, the output pulse at the OUT 2 pin, or pin 4, goes high and, in turn, resets the counter. Consequently, the counter is reset every two input cycles, giving a simple divide-by-2 circuit.

Step 4

Disconnect the power from the breadboard. Now connect pin 15 to pin 7 instead of pin 4. Apply power to the breadboard. What is the output frequency at pin 3 now?

Table 6-8. Record Output Frequency

RESET Connected To	Output Frequency	Divide-by-N
OUT 2 (pin 4)	5.000 kHz	2
OUT 3 (pin 7)	3.333 kHz	3
OUT 4 (pin 10)		
OUT 5 (pin 1)		
OUT 6 (pin 5)		
OUT 7 (pin 6)		
OUT 8 (pin 9)		
OUT 9 (pin 11)		

The output frequency should be 3.333 kHz, or one-third the input frequency. By connecting the RESET of the counter to the OUT 3 pin, the 4017 counter is now reset every three input cycles, giving a divide-by-3 circuit.

Step 5

Measure the output frequency by sequentially connecting the RESET input (pin 15) to the decoded outputs listed in Table 6-8.

You should have found that it is possible to divide an input frequency by any integer from 2 to 10 using the 4017 counter.

EXPERIMENT 3

Purpose

The purpose of this experiment is to demonstrate the operation of the 40192/74C192 programmable up/down counter.

Step 1

Wire the circuit as shown in the schematic diagram, setting the four logic switches all at logic 0. If you do not have a CMOS compatible 7-segment display, use four LED monitors instead, reading the output in binary (BCD) code.

Step 2

Apply power to the breadboard. What number do you see on the 7-segment display?

Pin Configuration of Integrated Circuit Chip (Fig. 6-48)

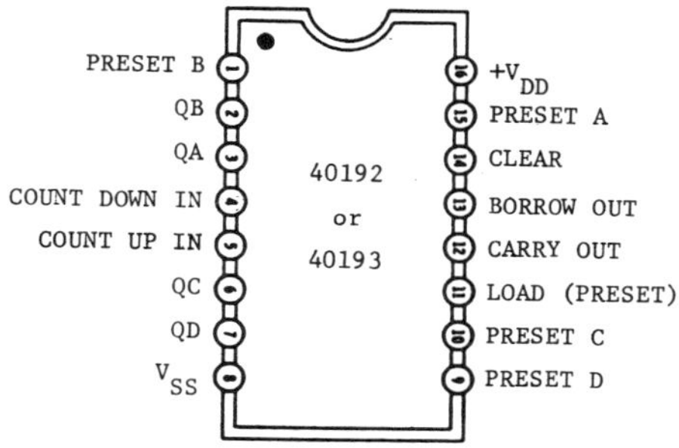

Fig. 6-48.

Schematic Diagram of Circuit (Fig. 6-49)

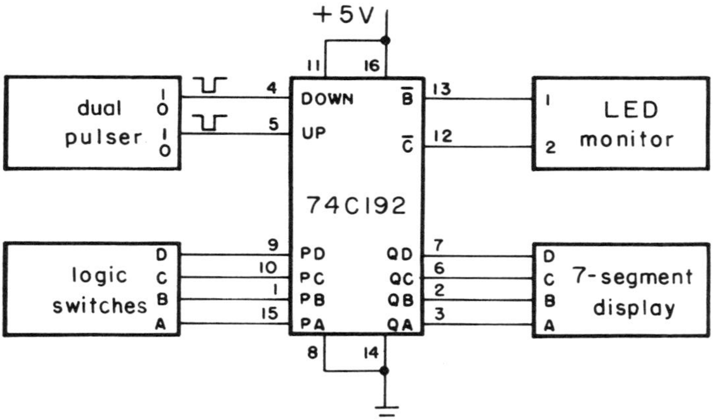

Fig. 6-49.

Depending on the initial state of the internal logic of the counter, any given number could appear when power is first applied. In our case, the number "0" was displayed. In addition, LED monitors No. 1 and No. 2 are lit, indicating that the $\overline{\text{BORROW}}$ and $\overline{\text{CARRY}}$ outputs of the counter are both at logic 1.

Step 3

Press and release the pulser that is connected to the UP input of the counter (pin 5) several times. What do you observe on the LED display as you press and release the pulser several times?

You should have observed that the LED display counts *upwards:* 0, 1, 2, 3, etc.

Step 4

Continue to press and release the pulser. What happens to the 7-segment display and LED monitor No. 2 after the count goes past "9"?

After the decade counter reaches its maximum count of 9, the $\overline{\text{CARRY}}$ signal, as indicated by LED monitor No. 2, goes out briefly, and when the display reads "0," LED monitor No. 2 lights up again. When several 40192 counters are cascaded, the $\overline{\text{CARRY}}$ signal is used as the "count up" input for the next decade.

Step 5

Continue to press and release the pulser until the number "6" appears on the LED display. Now press and release the pulser that is connected to the DOWN input of the counter (pin 4) three times. What do you observe on the LED display?

You should have observed that the LED display counts *downwards:* 6, 5, 4, 3.

Step 6

Press and release the pulser five more times. What happens to the 7-segment display and LED monitor No. 1 after the count goes past "0"?

After the counter reaches its minimum count of 0, the $\overline{\text{BORROW}}$ signal, as indicated by LED monitor No. 1, goes out briefly, and when the display then reads "9," LED monitor No. 1 lights up again. When several 40192 counters are cascaded, the $\overline{\text{BORROW}}$ signal is used as the "count down" input for the next decade.

Step 7

Remove the power from the breadboard. Now connect pin 11 to pin 13 instead of to the positive supply voltage. In addition, remove the following connections:

- Pulser to pin 5
- Pin 13 to LED monitor No. 1
- Pin 12 to LED monitor No. 2

Set the logic switches to the following BCD code: DCBA = 0111, and apply power to the breadboard. What number do you see on the display?

Again, depending on the state of the internal logic of the counter, any given number could appear when power is first applied. In our case, the number "4" was displayed.

Step 8

Now press the pulser several times until the display reads "0." Press the pulser again. What number now appears on the display?

When the pulser is pressed, the display should briefly read "7," and then read "6" when the pulser is released. When the count down goes past zero, the counter is parallel loaded with the 4-bit BCD number that is present at the inputs connected to the logic switches (0111 = 7). When the display shows the number "6," it then takes 7 input clock cycles (6, 5, 4, 3, 2, 1, and 0) until the counter is again loaded with the 4-bit BCD number.

Step 9

Now disconnect the power from the breadboard and connect a 1-kHz clock input to pin 4. Apply power to the breadboard and turn on the 1-kHz function generator. Set the logic switches at DCBA = 0001. What frequency do you observe?

Since the logic switches are set at DCBA = 0001, or the decimal number 1, the output frequency is the same as the input, or 1 kHz.

Step 10

Now vary the logic switch setting according to Table 6-9, and record the output frequency for each setting.

From your measurements, you should have observed that the output frequency is that fraction of the 1-kHz input set by the decimal equivalent of the 4-bit BCD code. We then have a 1-decade

Table 6-9. Record Output Frequency

D	C	B	A	N	Frequency
0	0	0	1	1	1000 Hz
1	0	0	0		
0	1	0	1		
0	0	1	1		
0	1	0	0		
1	0	0	1		
0	1	1	1		

programmable counter that is capable of dividing an input frequency by any integer from 1 to 9.

Step 11

What is the output frequency when you set the logic switches at DCBA = 0000?

Since the division by 0 is not defined mathematically, the 40192 or 74C192 counter is disabled so that the output frequency is the same as the input, or 1 kHz.

EXPERIMENT 4

Purpose

The purpose of this experiment is to demonstrate the operation of the Hughes HCTR0320 as a programmable divider.

Design Basics

- Binary modulus: $0 \leqslant N_B \leqslant 127$
- BCD (decimal) modulus: $0 \leqslant N_D \leqslant 999$
- Total modulus: $3 \leqslant N_D + N_B \leqslant 1023$

Step 1

The first point that should be emphasized is that the HCTR0320 is a relatively expensive device, costing approximately $10.00 to $15.00. Therefore, you are urged to be very careful when you use it. With both the power and the 1-kHz frequency reference discon-

Pin Configuration of Integrated Circuit Chip (Fig. 6-50)

Fig. 6-50.

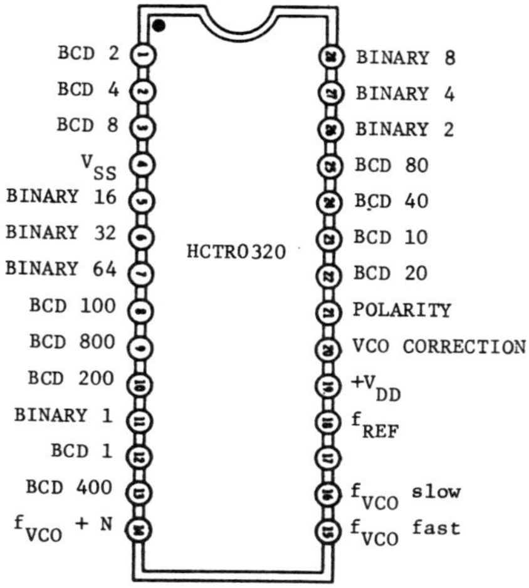

Schematic Diagram of Circuit (Fig. 6-51)

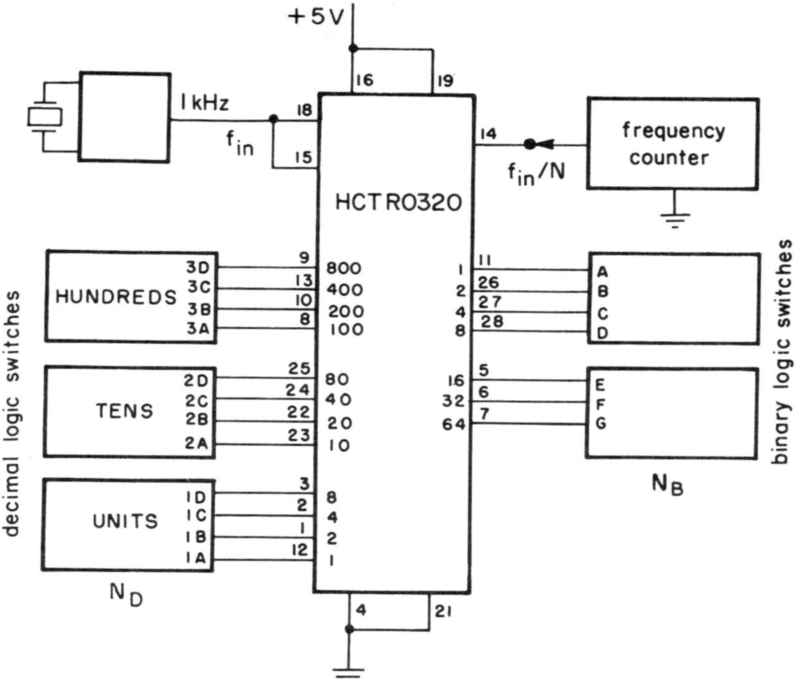

Fig. 6-51.

Table 6-10. Record Output Frequency

G	F	E	D	C	B	A	Output Frequency	Modulus
0	0	0	0	1	0	0	250 Hz	4
0	0	0	1	0	1	0		
0	0	1	0	1	0	0		
0	1	0	0	0	1	1		
0	1	0	1	0	0	0		
1	0	0	0	0	0	1		
1	1	0	0	1	0	0		
1	1	1	1	1	0	1		

nected from the breadboard, wire the circuit shown in the schematic diagram. If you do not have enough logic switches, use jumper wires to connect the programming inputs either to +5 volts (logic 1) or ground (logic 0) as instructed.

Step 2

Initially set the *binary* logic switches at GFEDCBA = 0000100, and all the *decimal* logic switches at logic 0. Then apply the power to the breadboard and connect the 1-kHz frequency reference to pins 15 and 18. Also connect the frequency counter to pin 14. What is the output frequency?

You should measure an output frequency of one-fourth the input, or 250 Hz, since the binary logic switches are set at 0000100, which is the binary equivalent of the number 4.

Step 3

Vary the binary logic switches according to Table 6-10 and record the measured output frequency and binary modulus.

For the last 7 settings, you should have measured frequencies of 200, 83, 50, 29, 25, 15, 10, and 8 Hz (±1 Hz), which correspond to divisions of 5, 12, 20, 35, 40, 65, 100, and 125, respectively.

Step 4

Now change the input frequency to 100 kHz. Set the *binary* logic switches at GFEDCBA = 0000111 (the decimal number 7). Also set the *units decimal* logic switches at the BCD code DCBA = 0011 (the decimal number 3); the *tens decimal* logic switches at

DCBA = 0111 (the decimal number 7); and the *hundreds decimal* logic switches at DCBA = 0000 (the decimal number 0). Consequently, the total decimal modulus is 073. To find the total modulus N, we add the binary modulus to the decimal modulus. For the logic switch settings in this step, what is the programmed modulus?

The total modulus is 80. Since the binary switches are set at 7, and the decimal logic switches are set at 73, the total modulus is 7 + 73, or 80. Now vary both the binary and decimal logic switches according to Table 6-11 and record the measured output frequencies. Note that the modulus is given as a decimal number. By this time you should be able to set the switches to the corresponding logic 0s and 1s.

You should have measured output frequencies that were 1/80, 1/4, 1/5, 1/10, 1/20, 1/250, 1/500, and 1/511 times the input frequency. If the input reference was exactly 100 kHz, then the output frequencies, to within 1 Hz, will be 1250 Hz, 25.0 kHz, 20.0 kHz, 10.0 kHz, 5.0 kHz, 400 Hz, 200 Hz, and 196 Hz.

Table 6-11. Record Output Frequency

Binary Modulus	Decimal Modulus			Total Modulus	Output Frequency
	Hundreds	Tens	Units		
7	0	7	3	80	
0	0	0	4	4	
1	0	0	4	5	
1	0	0	9	10	
9	0	1	1	20	
2	2	4	8	250	
4	4	9	6	500	
15	4	9	6	511	

EXPERIMENT 5

Purpose

The purpose of this experiment is to demonstrate the design and operation of a sine-wave generator made from a 4018 5-stage Johnson counter.

Pin Configuration of Integrated Circuit Chip (Fig. 6-52)

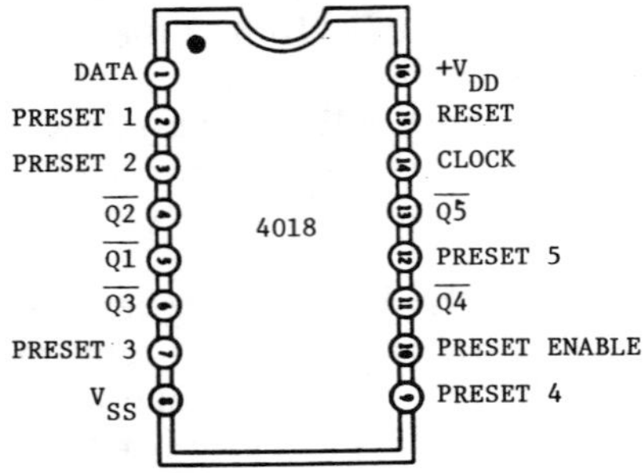

Fig. 6-52.

Schematic Diagram of Circuit (Fig. 6-53)

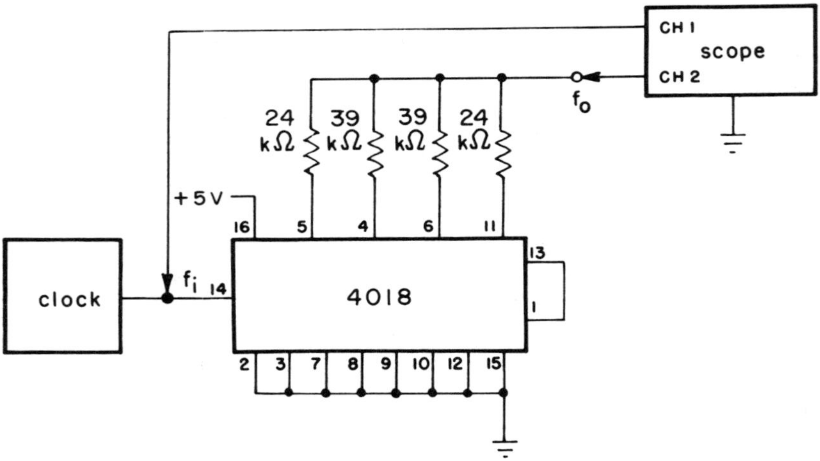

Fig. 6-53.

Design Basics

For a 5-stage counter:

- Output frequency: $f_o = f_i/10$
- Weighting resistors: $R_2/R_1 = 1.618$

$$R_3/R_1 = 1.618$$
$$R_4/R_1 = 1.000$$

Step 1

Wire the circuit shown in the schematic diagram and apply power to the breadboard. Next, set your oscilloscope to the following settings:

- Channel 1: 5 V/division
- Channel 2: 2 V/division
- Time base: 1 ms/division
- Triggering: Channel 2 (dc coupling)

Step 2

Position the input waveform (Channel 1) so that it is above the output waveform (Channel 2). Adjust the input square wave so that there are 10 complete cycles for the 10 horizontal scale divisions. You should notice that the output waveform, although a "chunky" sine wave, has a frequency that is one-tenth the input, since the 4018 is a 5-stage counter. For an *N*-stage counter, the output frequency will be 1/2N times the input.

Step 3

Now change the time base to 2 ms/division in order to display two full cycles of the output. Your output waveform should look like that shown in Fig. 6-54.

Step 4

Disconnect the power from the breadboard and add the simple passive low-pass filter shown in Fig. 6-55 to the output of the gen-

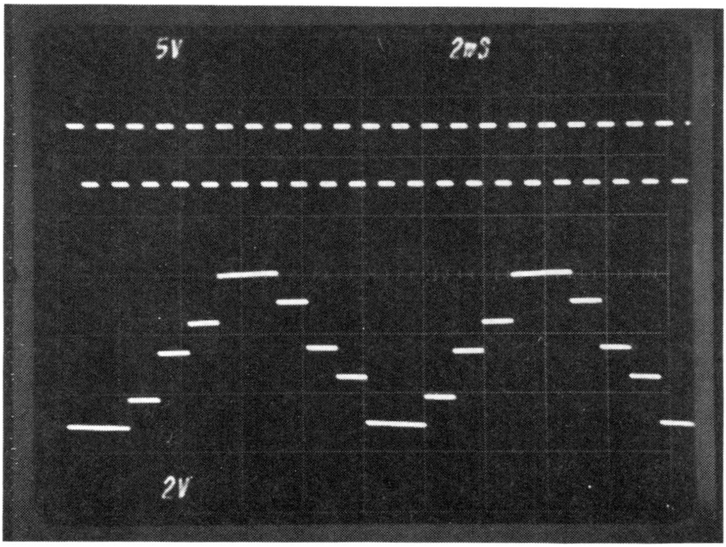

Fig. 6-54.

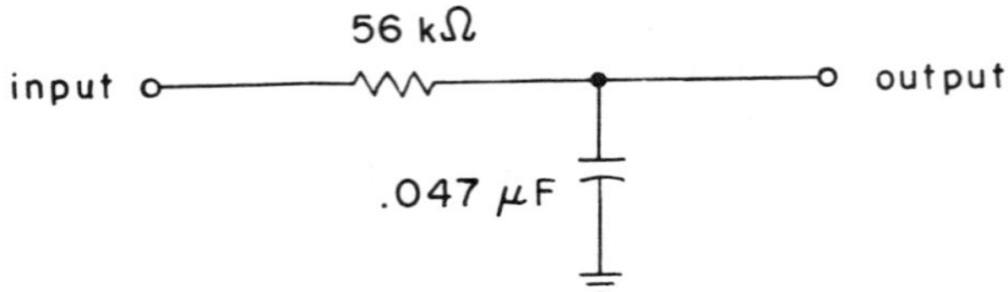

Fig. 6-55.

erator. Connect the Channel 1 probe of the oscilloscope to the input of the filter and the Channel 2 probe to the output of the filter. Connect the power to the breadboard. How does the output waveform compare with the waveform at the input of the filter?

The output waveform should resemble a cleaner-looking sine wave than the input. The low-pass filter removes the higher order harmonic components of the "chunky" waveform, leaving only the fundamental sine wave, as shown in Fig. 6-56.

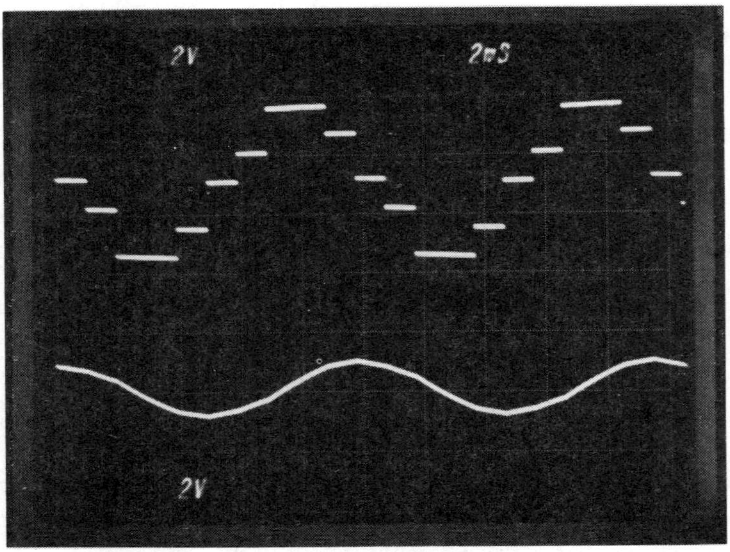

Fig. 6-56.

CHAPTER 7

Interfacing LEDs and 7-Segment LED Displays

INTRODUCTION AND OBJECTIVES

As with TTL systems, the design and operation of CMOS circuitry may be enhanced by using either colored LEDs or 7-segment LED numerical displays. However, because of the limited output drive associated with most CMOS gates and counters, the majority of LED displays are not able to be directly connected to the outputs of CMOS devices without some additional circuitry.

At the completion of this chapter, you will be able to do the following:

- Describe several methods for interfacing a single LED with a CMOS device.
- Interface common-anode and common-cathode 7-segment displays using CA3081 and CA3082 transistor arrays.
- Multiplex 7-segment displays.
- Convert 7-segment code to BCD using the 74C915 decoder.

INTERFACING LEDS

As shown in Fig. 7-1, the typical LED has a voltage drop of approximately 1.7 volts when forward biased. In addition, the amount of light emitted, which may be red, orange, yellow, or green in color, is proportional to the forward current I_F. Typically, the LED forward current ranges from 10 to 20 mA (50% to 100% relative brightness) set by a series current limiting resistor so that

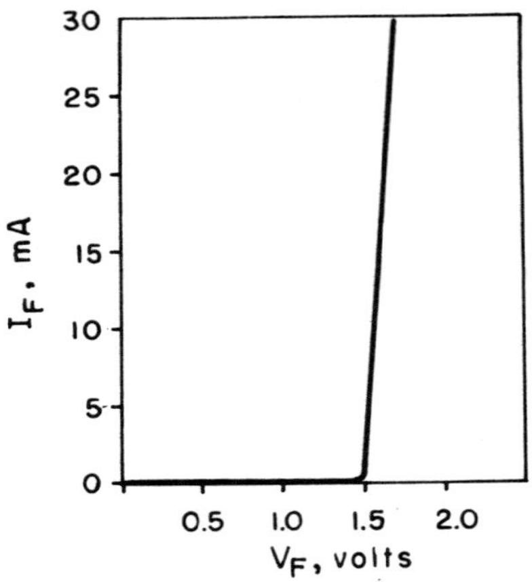

(A) Voltage drop across LED when forward biased.

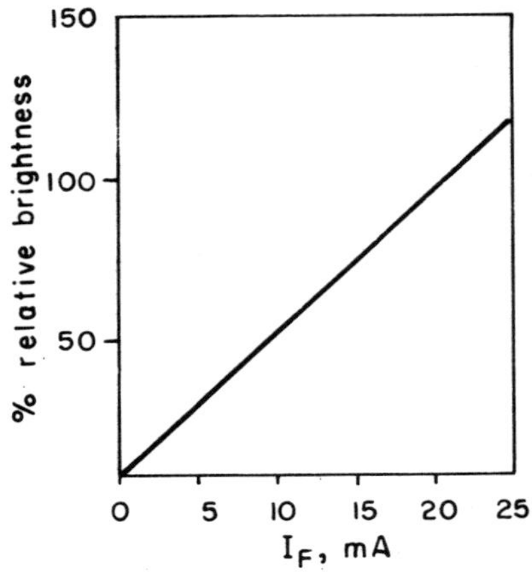

(B) Amount of light emitted is proportional to the forward current.

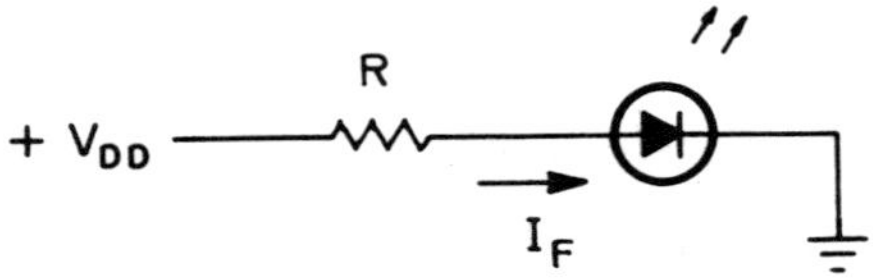

(C) Circuit.

Fig. 7-1. LED characteristics.

$$I_F = \frac{V_{DD} - V_F}{R} \quad \text{(Eq. 7-1)}$$

Unfortunately, most CMOS devices, because of their limited output current, cannot drive LEDs directly.

Fig. 7-2 illustrates several practical methods for driving LEDs. Figs. 7-2A and B use a transistor to help drive the LED, and the value for the current limiting resistor is essentially determined by using Equation 7-1. For either circuit, a logic 1 level at the base of the transistor causes the LED to light. Also, LEDs may be driven from the output of a 4050 CMOS buffer, as shown in Fig. 7-2C.

INTERFACING 7-SEGMENT LED DISPLAYS

Perhaps the easiest method to enable most CMOS counters to drive a wide variety of 7-segment LED displays is to use an integrated-circuit transistor array, such as the CA3081 or the CA3082, as shown in Fig. 7-3. The CA3081 consists of a series of seven high-current transistors in a common-emitter configuration for use with *common-anode* displays, while the CA3082 is a common-collector

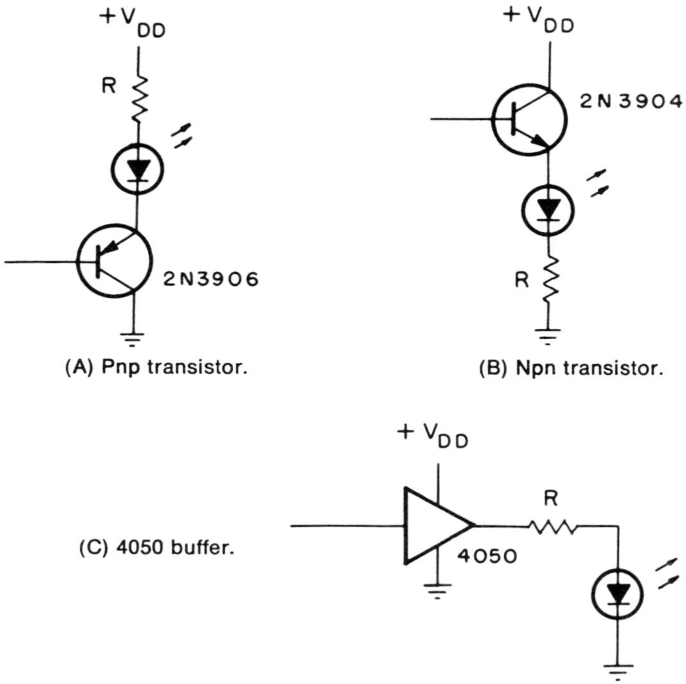

(A) Pnp transistor.

(B) Npn transistor.

(C) 4050 buffer.

Fig. 7-2. CMOS-to-LED interfaces.

array used with common-cathode displays. The following illustrates how 7-segment displays are interfaced with a variety of CMOS counters and decoders.

4026 and 4033 Counters

Fig. 7-4 shows how a MAN-3 common-cathode display is interfaced to either a 4026 or 4033 for supply voltages between 9 and

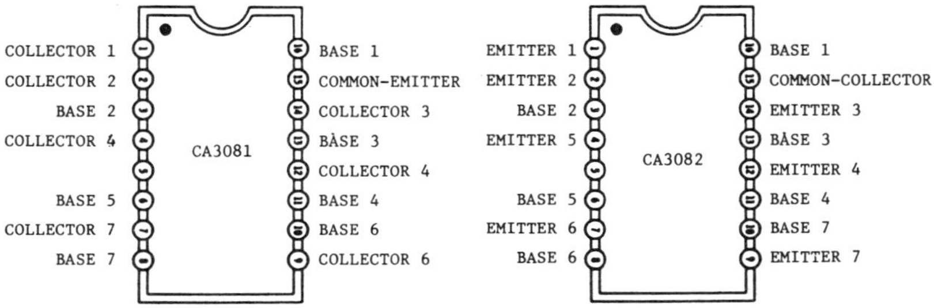

(A) CA3081 common-emitter-array.

(B) CA3082 common-collector-array.

Fig. 7-3. 7-Transistor arrays.

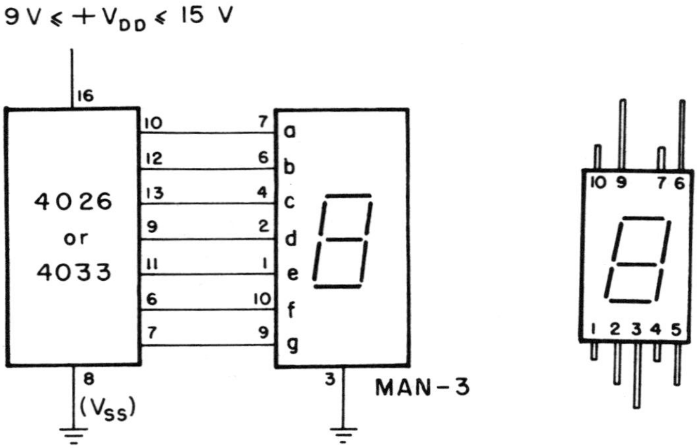

(A) Circuit.

(B) MAN-3 pin diagram.

Fig. 7-4. Interfacing a MAN-3 common-cathode LED display to either a 4026 or 4033.

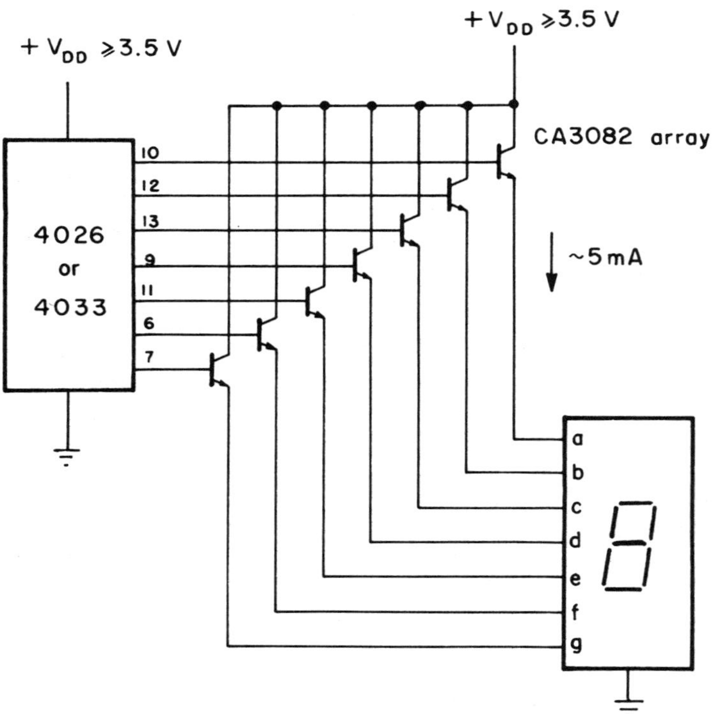

Fig. 7-5. Using a CA3082 array.

15 volts. This is, perhaps, the only case in which a CMOS device is able to drive a display directly without any additional buffering.

However, for supply voltages as low as 3.5 volts, a CA3082 common-collector array is required, as shown in Fig. 7-5. If a more precise control of the LED forward current is needed, a resistor in series with each LED segment should be added, as shown in Fig. 7-6. This emitter resistor is determined from:

$$R_E = \frac{V_{DD} - 2.5}{I_F} \qquad \text{(Eq. 7-2)}$$

The 2.5 volts in the numerator of Equation 7-3 collectively is the base-to-emitter voltage drop of the transistor (0.8 volt) plus the forward voltage drop of the LED (1.7 volts).

Also, a common-anode display, such as a MAN-7, can be used with the 4026 or 4033 counters by using a CA3081 common-emitter array as shown in Fig. 7-7. The collector resistor R_C for each segment is determined from:

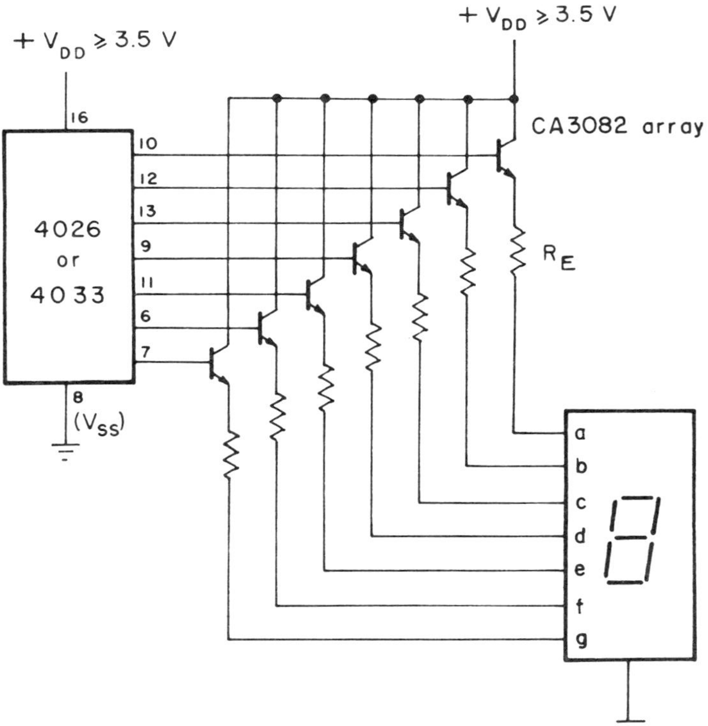

Fig. 7-6. Controlling the LED forward current.

$$R_C = \frac{V_{DD} - 1.9}{I_F} \qquad \text{(Eq. 7-3)}$$

where the 1.9-volt term in the numerator takes into account the forward voltage drop of the LED (1.7 volts) and the collector-to-

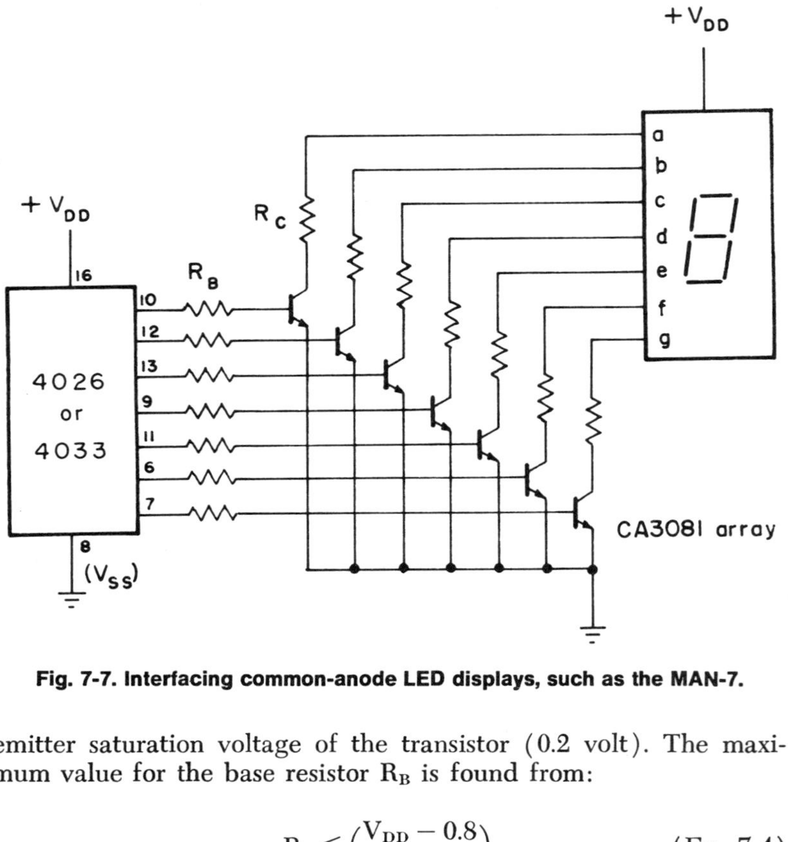

Fig. 7-7. Interfacing common-anode LED displays, such as the MAN-7.

emitter saturation voltage of the transistor (0.2 volt). The maximum value for the base resistor R_B is found from:

$$R_B \leqslant \left(\frac{V_{DD} - 0.8}{I_C/\beta_{min}}\right) \qquad \text{(Eq. 7-4)}$$

For both the CA3081 and CA3082 arrays, the minimum value for beta (β) is given as 30.

For both the 4026 and 4033 counters, the decoded output for the numbers 6 and 9 differs from that produced by the 7447 TTL decoder/driver. As illustrated in Fig. 7-8, the decoded output produces a "tail" on the numbers 6 and 9. For the number 6, the "a" segment is added, while the "d" segment is added for the number 9.

Fig. 7-8. Decoded numbers 6 and 9 with "tail."

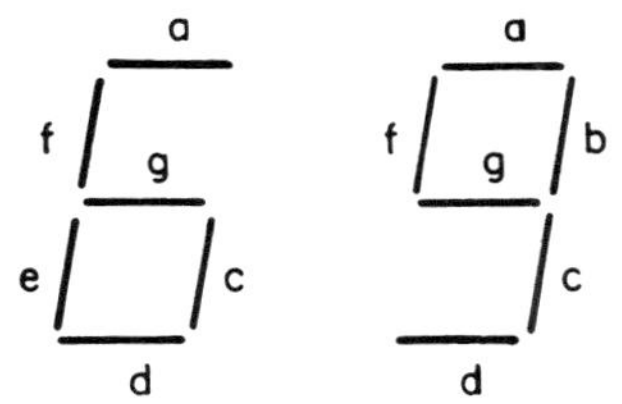

4511 7-Segment Latch and Decoder/Driver

The 4511 is constructed with npn bipolar transistor output drivers which are capable of sourcing up to 25 mA for each segment. Consequently, this amount of drive allows the 4511 to drive most common-cathode LED displays directly without buffering. Fig. 7-9 shows how a 4511 is connected to a DL704 common-cathode display. For each segment a current limiting resistor is required, and is determined from Equation 7-1. If the supply voltage is 9 volts and the LED forward current is to be 15 mA, then the series resistor is

$$R = \frac{V_{DD} - V_F}{I_F}$$

$$= \frac{(9 - 1.7)\ \text{volts}}{15\ \text{mA}}$$

$$= 487\ \Omega$$

for which a 470 ohm resistor can be used. The power dissipation of the resistor should also be determined. For a calculated value of 487 Ω and a current of 15 mA flowing through it, then by Ohm's law,

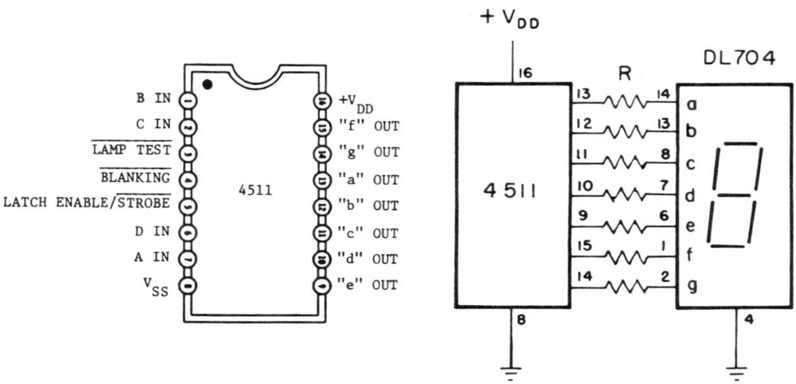

(A) Pin diagram.

(B) Interface with common-cathode LED display without buffering.

Fig. 7-9. 4511 7-segment latch and decoder/driver.

$$P = I^2R \quad \text{(Eq. 7-5)}$$
$$= (15\text{ mA})^2(487\ \Omega)$$
$$= 0.11\text{ watt}$$

so that a 1/4 watt resistor can be used safely.

Although the 4511 is primarily intended for common-cathode displays, a common-anode display similar to the MAN-1 or MAN-7 can be used with a CA3081 transistor array as shown in Fig. 7-10.

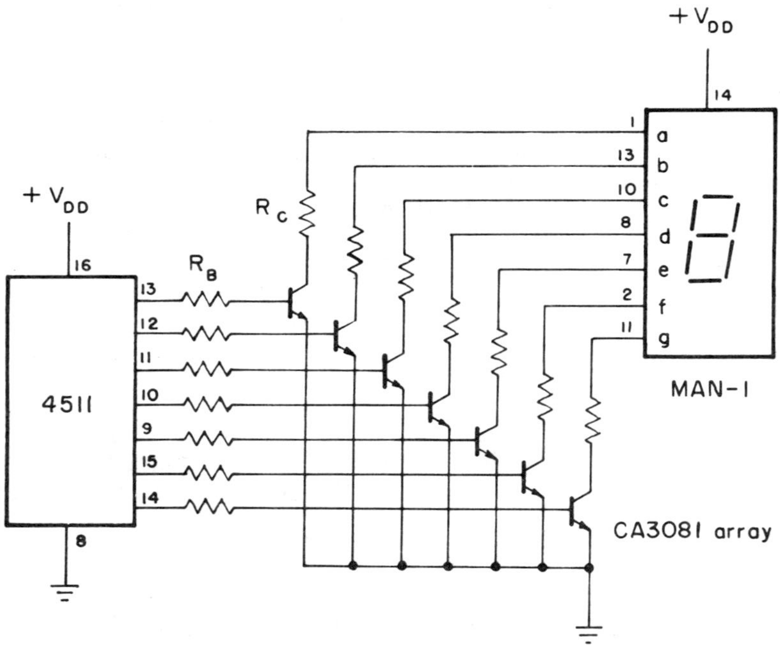

Fig. 7-10. The 4511 decoder/driver interfaced with common-anode LED display.

MULTIPLEXED 7-SEGMENT DISPLAYS

For LED displays requiring 4 or more digits, it would probably be more advantageous to multiplex, or strobe, the display. In general, display multiplexing is a rather efficient method that uses a single decoder/driver circuit for all the digits as illustrated in block form in Fig. 7-11. The single decoder/driver is sequentially time shared with each of the LED displays, one display at a time. Since

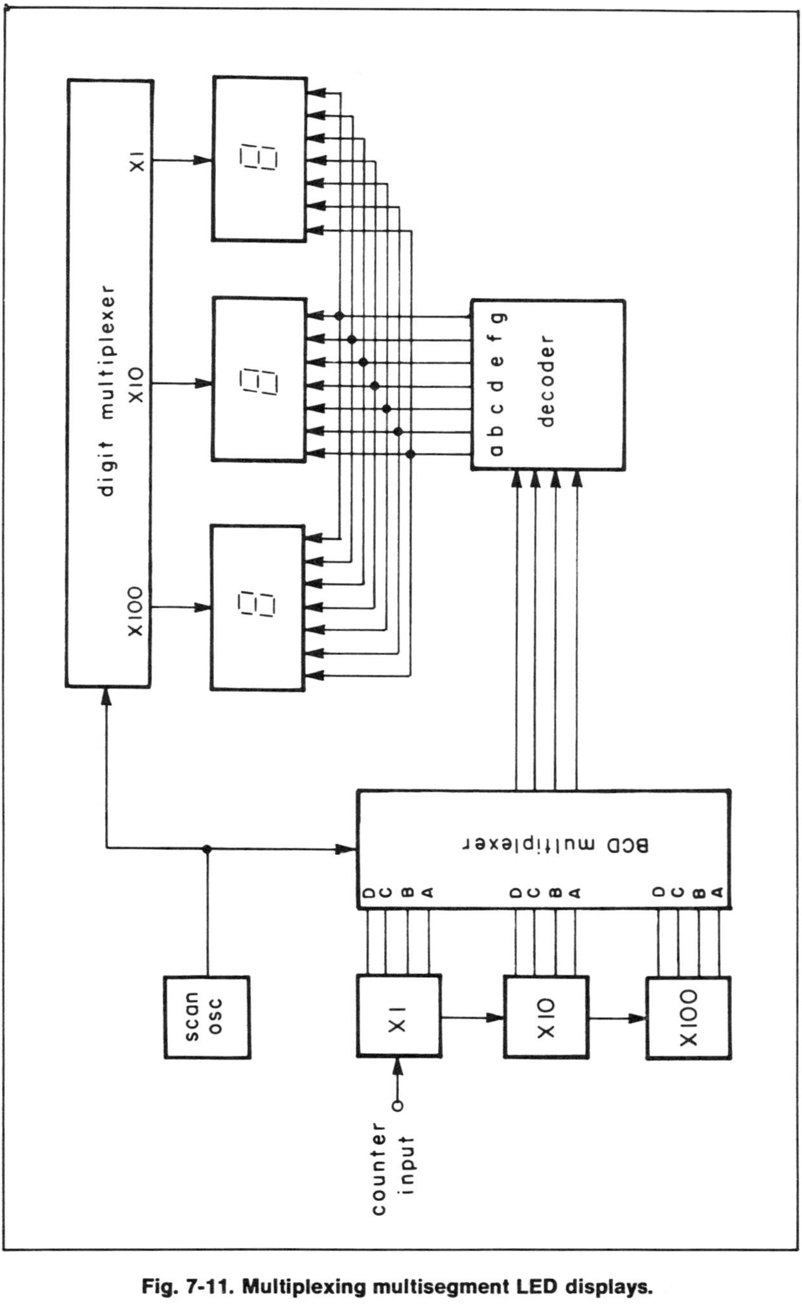

Fig. 7-11. Multiplexing multisegment LED displays.

all of the like segments are wired in parallel, each digit, out of *N* total digits, will be enabled for 1/*N* the time it takes to strobe all *N* digits.

If the display is multiplexed without any increase in the forward current to each of the seven LED segments, the resultant display will be dim. This is because the *average* current will now be less than that without multiplexed displays, so that the following equation will be:

$$I_F \text{ (average)} = \frac{I_F}{N} \quad \text{(Eq. 7-6)}$$

For a MAN-7 display normally operating with a forward current of 15 mA, the average forward current for each segment of 4 multiplexed displays will then be 15 mA/4, or 3.75 mA. By increasing the forward current to *N times* the normal value, the average value will then be sufficient to have a visible display. In the above example, an average forward current of 15 mA results if the peak current is set at 4 times 15 mA, or 60 mA, for a 4-digit multiplexed display. However, when the individual LED display is operated in a pulsed mode and the peak forward current is increased in order to increase the average current, the maximum allowable duty cycle must not be exceeded or the individual diodes of the display will be destroyed.

The *scan rate,* or strobing frequency must be high enough so that there is no noticeable flicker as each digit is sequentially enabled. The human eye can distinguish light pulses up to about 16 Hz. Above this, the light pulses tend to appear as a continuous glow, but vary somewhat between individuals, viewing, and ambient lighting conditions.

Using a series of 4033 counters and MAN-3 common-cathode LED displays, Fig. 7-12 illustrates the method of properly multiplexing *N* digits. A series of *N* CA3082 common-collector arrays are used to drive the display segments, while a CA3081 common-emitter array is used to enable the common-cathode terminal of each digit as the entire display is scanned. Each segment of the CA3081 array is connected to a divide-by-N decoder, such as the 4017 and 4022 devices. Fig. 7-13 shows a similar scheme, but with a series of 4026 counters. This method enables the individual counter instead of switching the common-cathode of the LED display to ground. Specific information on how the 4017 and 4022 counters are connected to divide/decode any number from 2 to 10 is found in Chapter 6.

Fig. 7-14 illustrates how a series of four 4026 counters are connected to a multidigit monolithic LED display, such as the 5082-7400 series manufactured by Hewlett-Packard.

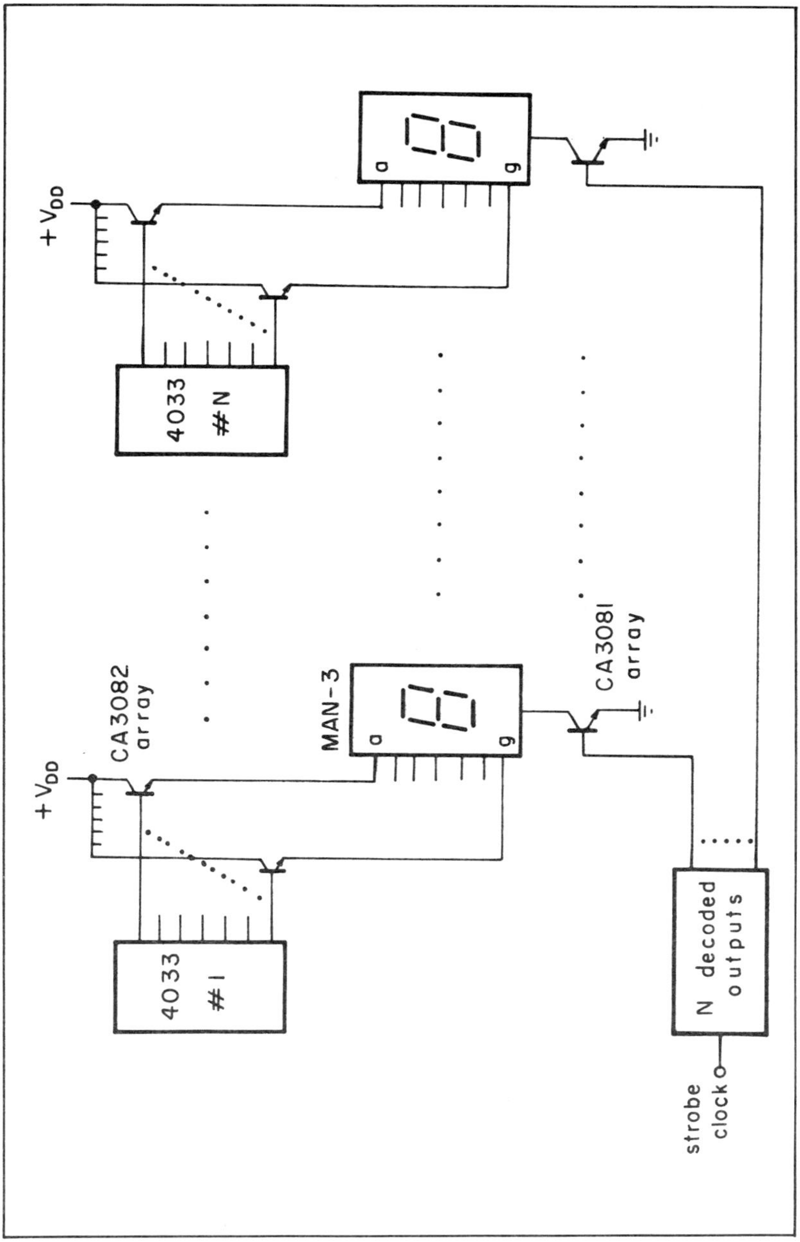

Fig. 7-12. Multiplexing a series of 4033 counters with common-cathode LED displays.

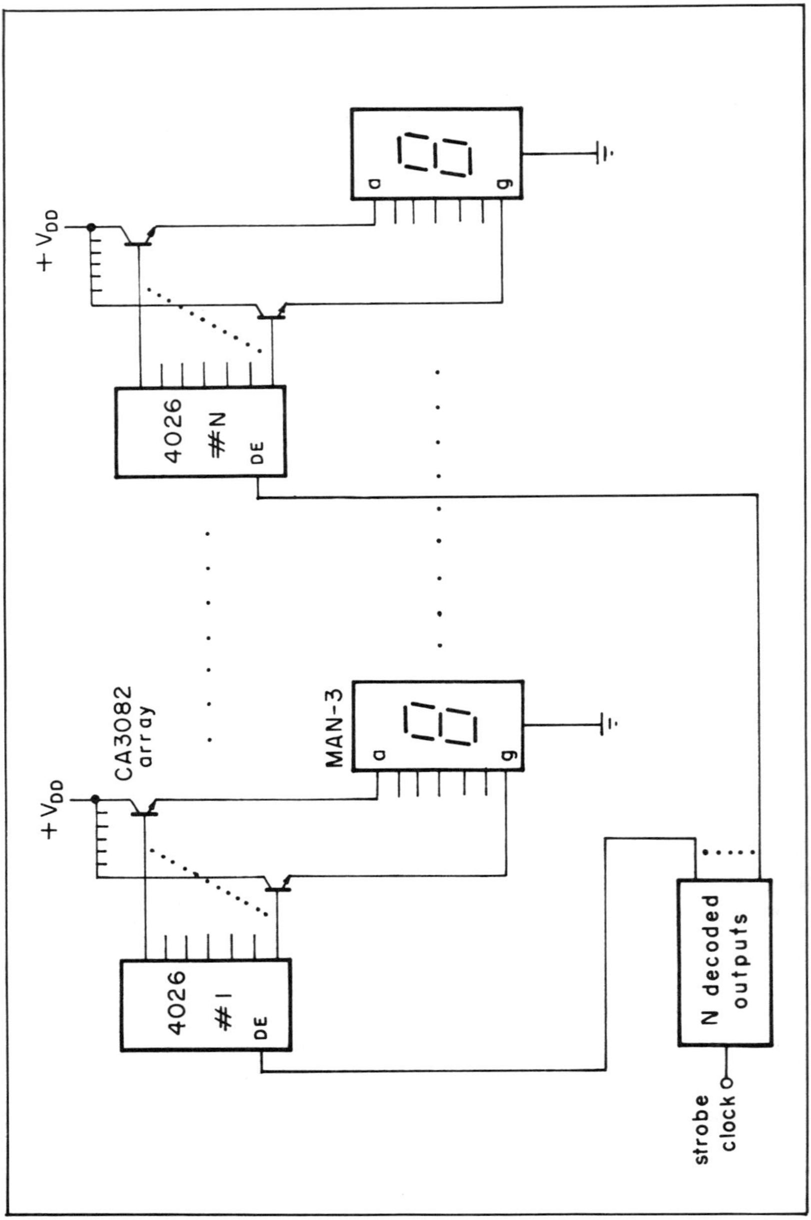

Fig. 7-13. Multiplexing a series of 4026 counters with common-cathode LED displays.

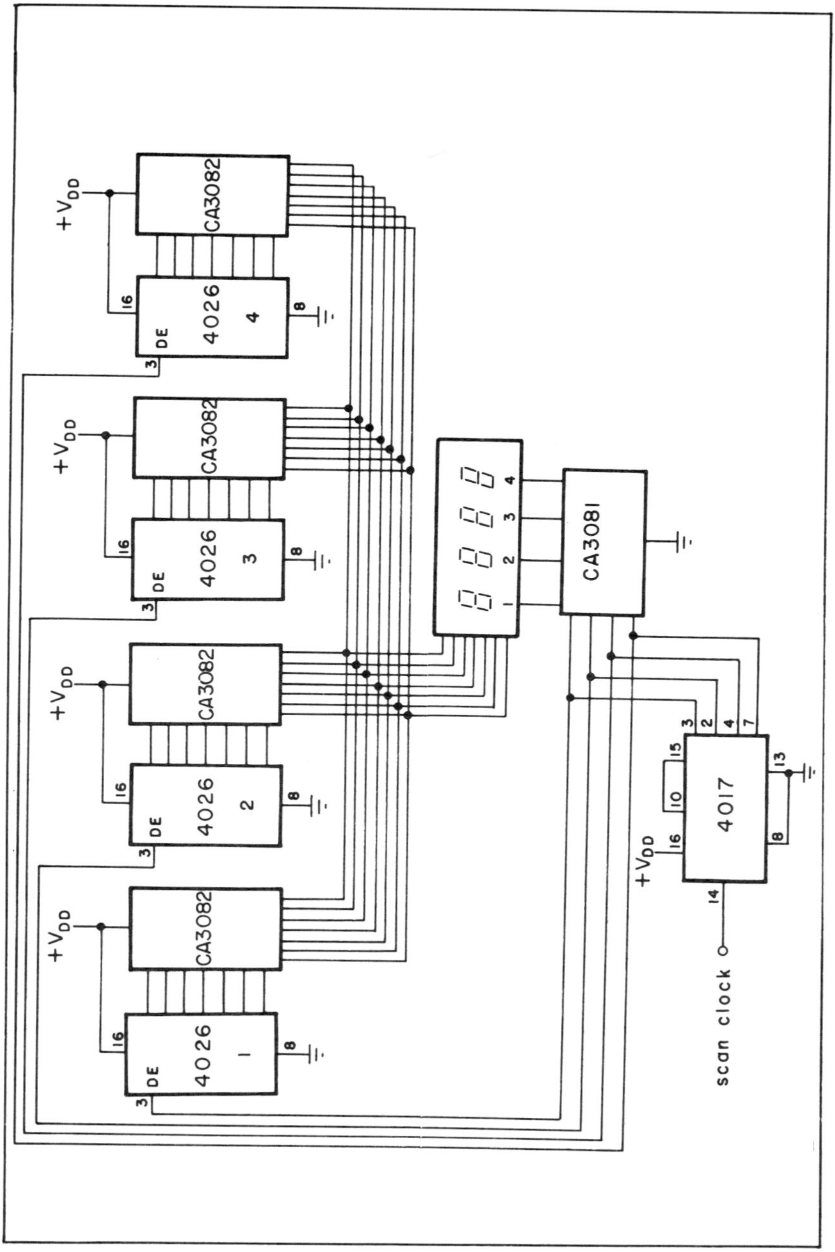

Fig. 7-14. Multiplexing four 4026 counters with a 4-digit LED display.

OBTAINING BCD INFORMATION FROM 7-SEGMENT DISPLAYS

For a number of applications, particularly microcomputer interfacing, it may be appropriate to obtain the equivalent 4-bit BCD code corresponding to the number shown on each digit of a series of 7-segment LED displays. Almost all CMOS-type digital clocks, panel meters, pocket calculators, and counters (4026 and 4033) have no provisions for obtaining BCD data.

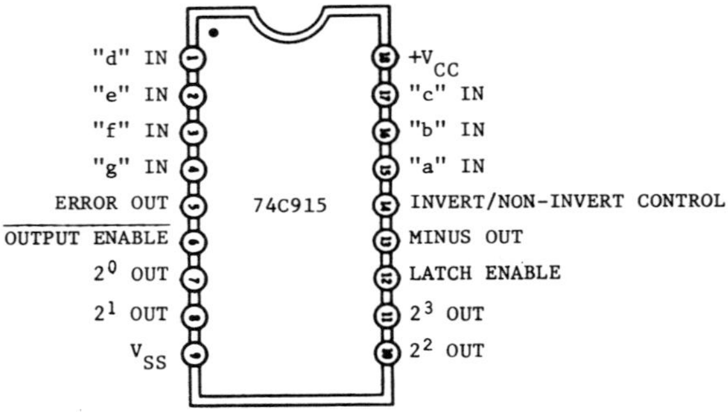

Fig. 7-15. Pin diagram of the 74C915 7-segment-to-BCD decoder.

With a 74C915 decoder (shown in Fig. 7-15 and manufactured by National Semiconductor), we are able to convert the 7-segment information (Table 7-1) into BCD code. The true state of the 7-segment inputs can be selected by usage of the INVERT/NONINVERT control input. A logic 0 on this control input selects active HIGH true decoding at the 7-segment inputs. The ERROR output is a logic 1 whenever a nonstandard 7-segment code is present at the 7 inputs. The MINUS output is a logic 1 whenever the minus code is detected. When the LATCH ENABLE (LE) input is at logic 1, the BCD output is latched (TRI-STATE outputs), and when at logic 0, the BCD outputs are unlatched.

AN INTRODUCTION TO THE EXPERIMENTS

Most CMOS devices, because of their limited output current, cannot drive LED displays directly. Consequently, a CMOS-to-LED interface is required for most cases, without destroying the CMOS device, the LED display, or both. The experiments that you will perform can be summarized as follows:

Table 7-1. 74C915 7-Segment/BCD Truth Table

7-Segment Character	Segments a	b	c	d	e	f	g	BCD Outputs D	C	B	A	ERROR Output	MINUS Output
0	1	1	1	1	1	1	0	0	0	0	0	0	0
1	0	1	1	0	0	0	0	0	0	0	1	0	0
1	0	0	0	0	1	1	0	0	0	0	1	0	0
2	1	1	0	1	1	0	1	0	0	1	0	0	0
3	1	1	1	1	0	0	1	0	0	1	1	0	0
4	0	1	1	0	0	1	1	0	1	0	0	0	0
5	1	0	1	1	0	1	1	0	1	0	1	0	0
6	1	0	1	1	1	1	1	0	1	1	0	0	0
6	0	0	1	1	1	1	1	0	1	1	0	0	0
7	1	1	1	0	0	0	0	0	1	1	1	0	0
8	1	1	1	1	1	1	1	1	0	0	0	0	0
9	1	1	1	1	0	1	1	1	0	0	1	0	0
9	1	1	1	0	0	1	1	1	0	0	1	0	0
blank	0	0	0	0	0	0	0	1	1	1	1	0	0
—	0	0	0	0	0	0	1	X	X	X	X	1	1

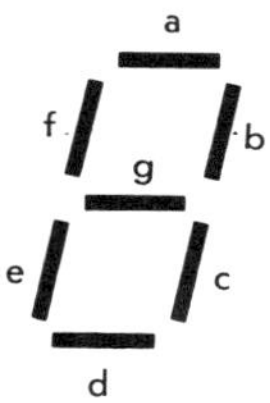

X = represents TRI-STATE condition

Experiment No.	*Purpose*
1	Demonstrates how LEDs are interfaced with CMOS devices, using either transistors or CMOS buffers.
2	Demonstrates how a common-cathode 7-segment LED display is connected to a 4511 BCD-to-7-segment decoder/driver.

EXPERIMENT 1

Purpose

The purpose of this experiment is to demonstrate how LEDs are properly interfaced with CMOS devices, using either an npn transistor or a 4050 CMOS buffer.

Pin Configuration of Integrated Circuit Chips (Fig. 7-16)

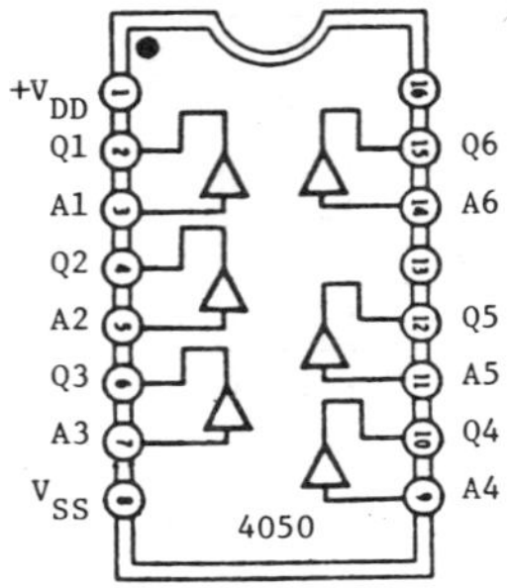

Fig. 7-16.

Schematic Diagram of Circuit (Fig. 7-17)

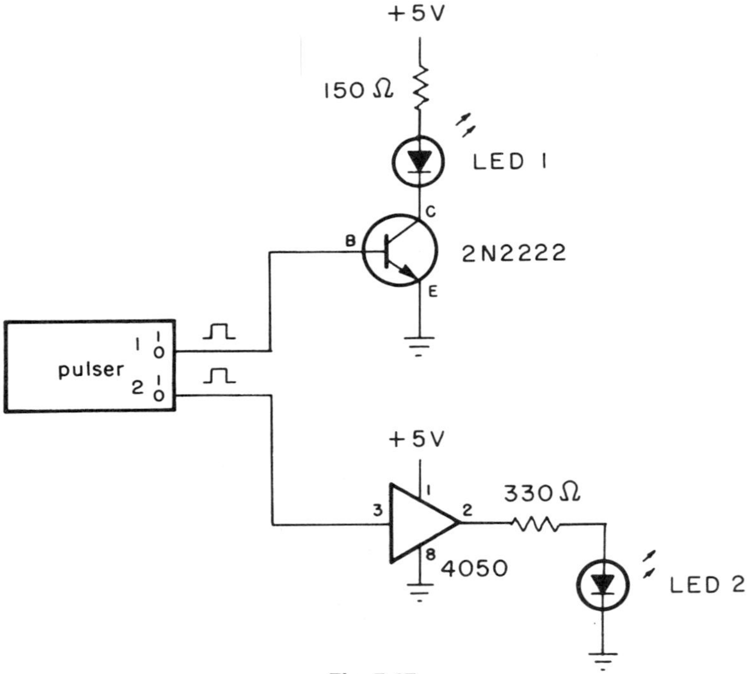

Fig. 7-17.

Step 1

Wire the circuit as shown in the schematic diagram. For this experiment, the two pulsers represent the CMOS circuit that is to be interfaced with the LEDs. Since the unused inputs of a CMOS device cannot be left floating, connect pins 5, 7, 9, 11, and 14 of the 4050 buffer to ground. In addition, make sure that the LEDs and the transistor are connected properly.

Step 2

Now connect the power to the breadboard. Is LED No. 1 lit or unlit?

LED No. 1 should be unlit, since the input at the base of the transistor is at logic 0.

Step 3

Press in pulser No. 1. What, if anything, happens to LED No. 1?

LED No. 1 should now be lit, since the output of the CMOS circuit (the pulser) is now at logic 1. When you release the pulser the LED should be unlit, as in Step 2.

Step 4

For the remaining circuit, is LED No. 2 lit or unlit?

LED No. 2 should be unlit, since the output of pulser No. 2 is presently at logic 0.

Step 5

Press in pulser No. 2. What, if anything, happens to LED No. 2?

LED No. 2 should now be lit, since the output of the CMOS circuit (the pulser) is now at logic 1. When you release the pulser, the LED should be unlit, as in Step 4.

EXPERIMENT 2

Purpose

The purpose of this experiment is to demonstrate the operation of the 4511 BCD-to-7-segment LED display decoder/latch/driver when interfaced with a DL704 common-cathode display.

Pin Configuration of Integrated Circuit Chip and LED Display (Fig. 7-18)

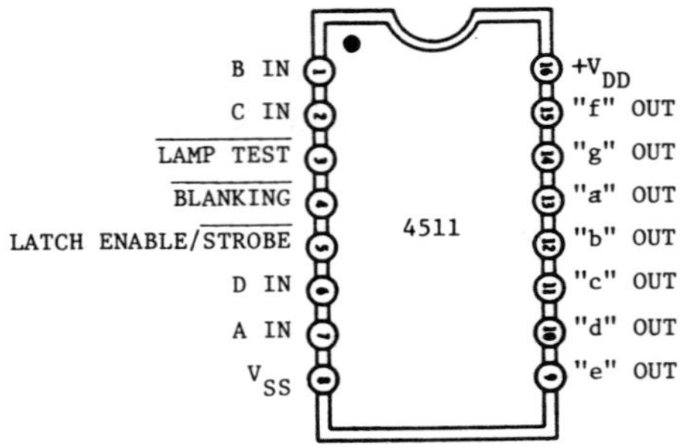

Fig. 7-18.

Schematic Diagram of Circuit (Fig. 7-19)

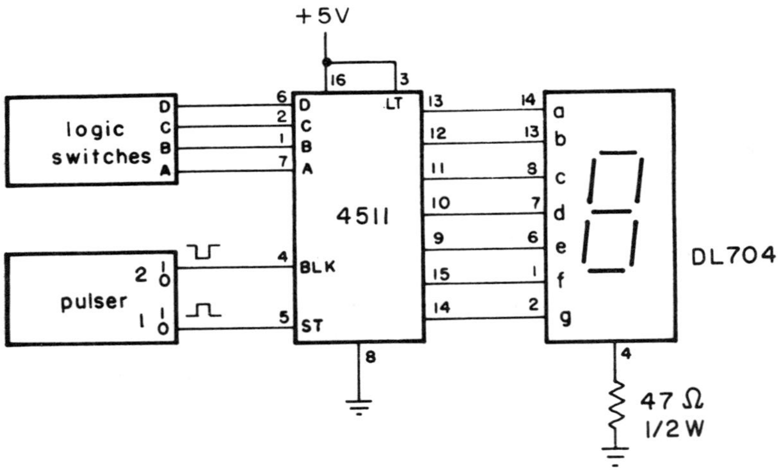

Fig. 7-19.

Step 1

Wire the circuit as shown in the schematic diagram and apply power to the breadboard. Set the four logic switches so that DCBA = 0000. What number appears on the LED display?

The number 0 should be displayed, since the 4511 decodes the 4-bit BCD input and the LED display shows this code as a decimal number. For BCD codes ranging from DCBA = 0001 (decimal 1) to DCBA = 1001 (decimal 9), verify that the number shown by the LED display is the equivalent of the 4-bit BCD code settings of the logic switches.

Step 2

What happens to the LED display when the logic switches are set at either DCBA = 1010, 1011, 1100, 1101, 1110, or 1111?

When the logic switches are set at those 4-bit BCD codes which are the decimal equivalents of the numbers 10, 11, 12, 13, 14, and 15, the display is blank, since it would make no sense for a BCD-to-7-segment display to indicate those 4-bit codes higher than the decimal number 9.

Step 3

Now set the logic switches so that DCBA = 0111, i.e., the decimal number 7. Press and hold the pulser (pulser No. 1) that is connected to the STROBE (latch) input of the 4511 at pin 5. While holding pulser No. 1 in, so that the STROBE input is a logic 1, change logic switch "A" to logic 0. Does anything happen on the LED display?

Nothing should have happened. Now release the pulser. Did anything happen then?

The display should now read the number 6, since the input code was changed from DCBA = 0111 to 0110 while the STROBE input was at logic 1. Perhaps a better name for this input should be "LATCH ENABLE" when this input is at logic 1.

Step 4

Press in pulser No. 2, which is connected to pin 4 of the 4511. What happened?

When this pulser is pressed in so that pin 4 is now at logic 0, the display is blanked! This feature is handy when you want to save power by turning off the display when not needed.

CHAPTER 8

Specialized CMOS Devices

INTRODUCTION AND OBJECTIVES

Like their TTL counterparts, CMOS devices such as gates, latches, and counters, etc., are easily interconnected to create entire systems: timers, digital time clocks, frequency/event counters, and digital panel meters to name a few. However, when such systems are built, the total number of integrated circuits required can be overwhelming.

Rather than reinvent the wheel each time, several CMOS manufacturers have now developed a number of useful single-chip systems, which are intended to replace a majority, if not all the discrete devices of a functionally equivalent circuit. This final chapter discusses several of these one-chip systems, such as phase-locked loops and synthesizers, multidigit counters/frequency counters, Touch-Tone encoders, oscillators/dividers, and 3 1/2 digit A/D converters.

At the completion of this chapter you will be able to do the following:

- Learn how to use single-chip systems to reduce the number of components with the following devices as examples:
  - MC14410 and ICM7206 Touch-Tone Encoders
  - MC14433 and ICL7106/7107 3 1/2 digit A/D converters
  - 4046 phase-locked loop
  - ICM7207/7208 two-chip frequency counter
  - 74C926 4-decade counter

TOUCH-TONE ENCODERS

Stable Touch-Tone* signals, used for telephone dialing and other communication services, can be encoded by several CMOS devices

* Touch-Tone is a registered trademark of American Telephone and Telegraph.

with readily available quartz crystals. Touch-Tone signals are coded in tone-pairs, using two of eight possible tone frequencies for the numbers 0 through 9, letters A through D, and the symbols # (pound) and * (star). The possible eight audio frequencies are shown in matrix form in Table 8-1, so that the tones are divided into a "low" and a "high" group. Representative CMOS encoders are the MC14410 and the ICM7206.

Table 8-1. Audio Frequencies Used for Touch-Tone Coding

Low Tone Group (Hz)	High Tone Group			
	1209	1336	1477	1633
697	1	2	3	A
770	4	5	6	B
852	7	8	9	C
941	*	0	#	D

The Motorola MC14410 (Fig. 8-1) permits the design of a low-power Touch-Tone encoder using a 1-MHz crystal, as shown in Fig. 8-2, and features multiple key lockout with a power consumption of approximately 15 mW at 5 volts during standby.

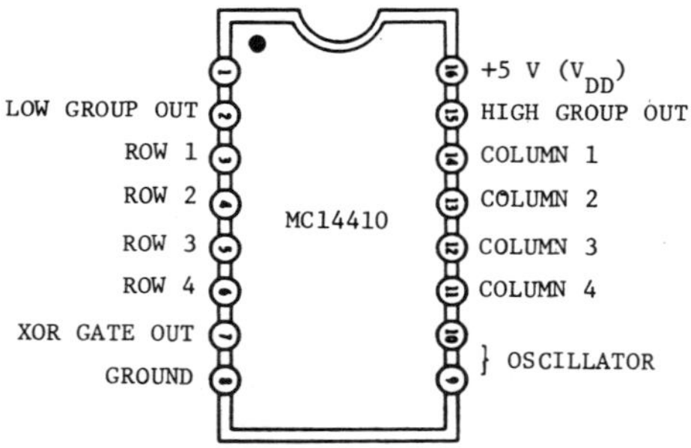

Fig. 8-1. MC14410 pin diagram.

The inputs from a 2-of-8 key pad, such as the Chromerics model ER-21611, originate from a 4 × 4 switch matrix generating a four row (R1 – R4) by four column (C1 – C4) input signal according to Table 8-2. As an example, if the number "7" is pressed on the pad, the "row 3" and "column 1" inputs are grounded which program the internal logic to simultaneously deliver an 852-Hz low tone and a 1209-Hz high tone. These two outputs are then summed together via resistors to give the resultant two-tone output. The output at pin 7 is a pulsed logic output, obtained by feeding the outputs of

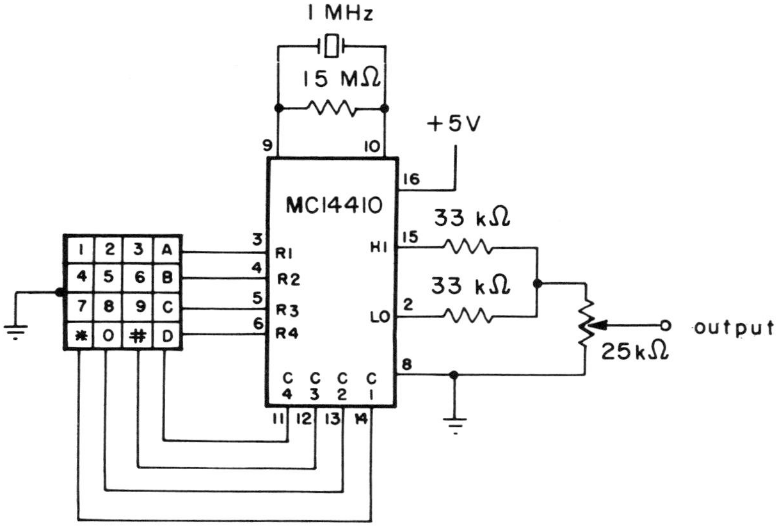

Fig. 8-2. MC14410 Touch-Tone encoder circuit.

the frequency dividers of the two-tone groups to an exclusive-NOR gate. This output can be used as a keyed push-to-talk signal to drive a monostable multivibrator when any of the keys are briefly pressed.

The Intersil ICM7206 (Fig. 8-3) is somewhat similar to the MC14410. It also consumes less than 8 mW and has multiple key lockout. However, the 7206 requires a 3.579545-MHz "color tv" crystal, as shown in the operating circuit of Fig. 8-4.

3 1/2 DIGIT ANALOG-TO-DIGITAL CONVERTERS

All digital panel meters, such as a digital voltmeter, use some form of analog-to-digital (A/D) conversion. With the increasing

Fig. 8-3. ICM7206 pin diagram.

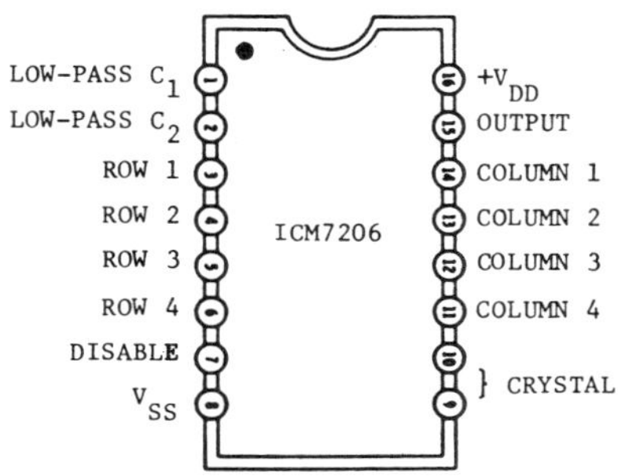

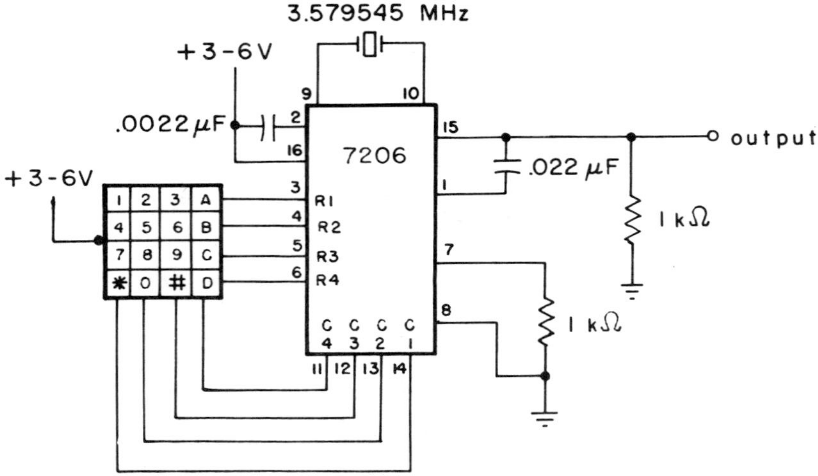

Fig. 8-4. ICM7206 Touch-Tone encoder circuit.

use of microprocessors for data acquisition, A/D converters are frequently used to input analog data. Among the popular CMOS devices presently available are Motorola's MC14433, and the ICL-7106 (LCD drive), and ICL7107 (LED drive) made by Intersil.

The MC14433, as shown by the pin diagram of Fig. 8-5, is a single-chip 3-1/2 digit A/D converter using a modified dual ramp technique of A/D conversion with a high input impedance, autopolarity, and auto-zero.

The digital output is a multiplexed 3-1/2 digit BCD code with the most significant digit containing over/underrange and polarity in-

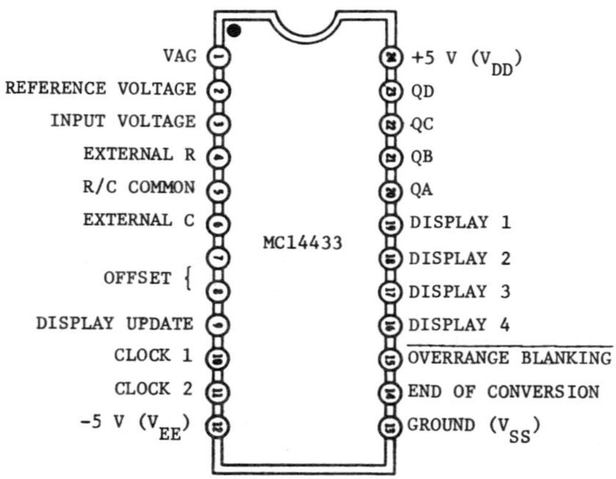

Fig. 8-5. MC14433 pin diagram.

Table 8-2. Touch-Tone Encoder Truth Table

Character Input	Rows				Columns				High-Tone Group	Low-Tone Group
	R1	R2	R3	R4	C1	C2	C3	C4		
1	0	1	1	1	0	1	1	1	1209	697
2	0	1	1	1	1	0	1	1	1336	697
3	0	1	1	1	1	1	0	1	1477	697
A	0	1	1	1	1	1	1	0	1633	697
4	1	0	1	1	0	1	1	1	1209	770
5	1	0	1	1	1	0	1	1	1336	770
6	1	0	1	1	1	1	0	1	1477	770
B	1	0	1	1	1	1	1	0	1633	770
7	1	1	0	1	0	1	1	1	1209	852
8	1	1	0	1	1	0	1	1	1336	852
9	1	1	0	1	1	1	0	1	1477	852
C	1	1	0	1	1	1	1	0	1633	852
*	1	1	1	0	0	1	1	1	1209	941
0	1	1	1	0	1	0	1	1	1336	941
#	1	1	1	0	1	1	0	1	1477	941
D	1	1	1	0	1	1	1	0	1633	941

formation, as well as the half digit (blank or "1"). The MC14433 can be externally set to display a full-scale voltage of either ±199.9 mV or ±1.999 V by the values listed in Table 8-3 for the reference voltage and integrator resistor. A pair of 0.1 μF capacitors are used for the integrator (pins 5 and 6) and for offset correction (pins 7 and 8). Although an external clock may be used at pin 10, the MC14433 nevertheless contains an internal clock (pin 11) so that a single resistor can be used between pins 10 and 11 to set the frequency of the conversion cycle, composed of the unknown input measurement and auto-zero cycles.

The END OF CONVERSION signal (pin 14) indicates the end of one conversion and the start of the next. In addition, the DISPLAY UPDATE signal at pin 9 permits data to be strobed to the output latches. In most cases these two pins are tied together, allowing the output to be updated at the end of each conversion cycle.

Fig. 8-6 shows how the MC14433 is used to create a simple digital voltmeter,* with a full-scale reading of ±1.999 volts. A 4511 decodes the 4-bit BCD output to the 3-1/2 digit common-cathode display.

---

* Ahrens, T. "Build a 3-1/2 Digit DVM," 73 *Magazine*, February, 1978, pp. 40–42.

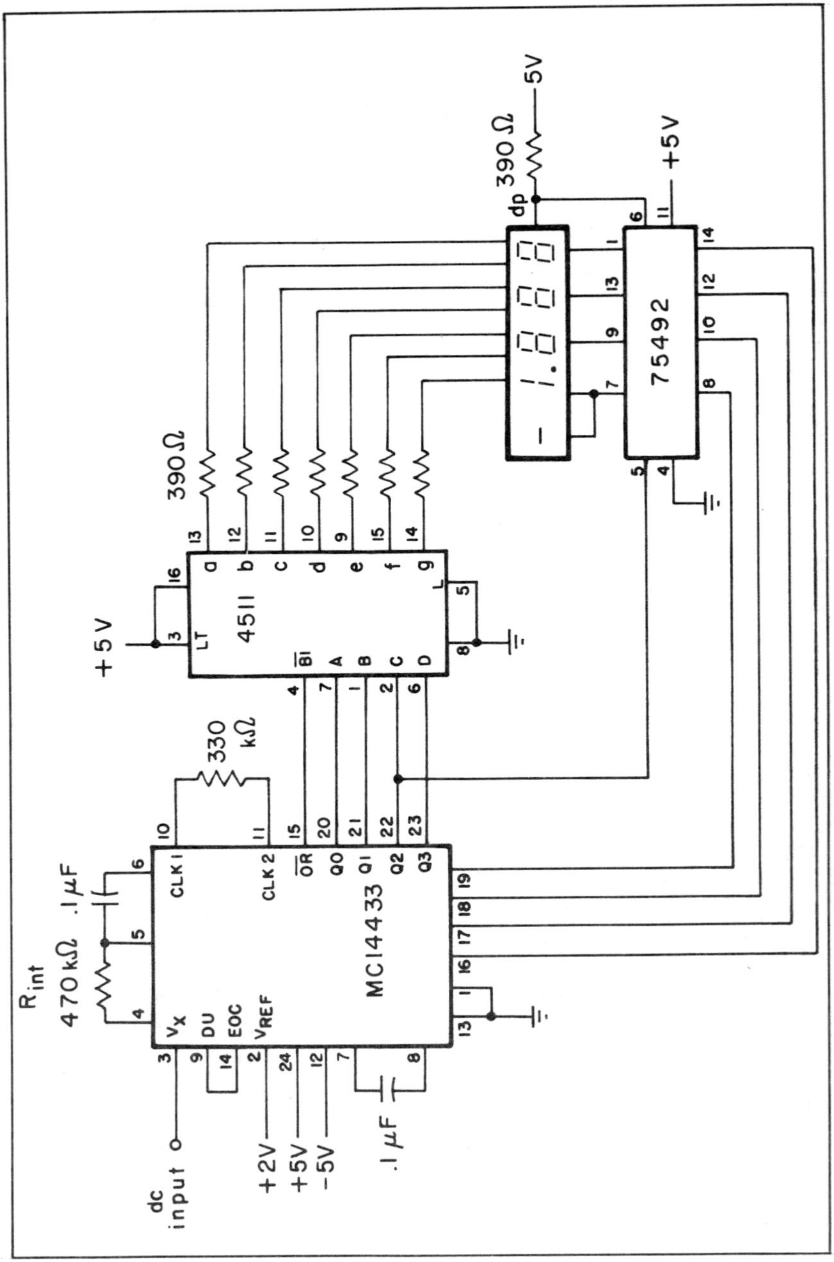

Fig. 8-6. Digital voltmeter.

Table 8-3. Reference Voltages and Integrator Resistors for Full-Scale Voltage

Full-Scale Voltage	Reference Voltage (pin 2)	Integrator Resistor (pins 4 & 5)
±199.9 mV	+200.0 mV	27 kΩ
±1.999 V	+2.000 V	470 kΩ

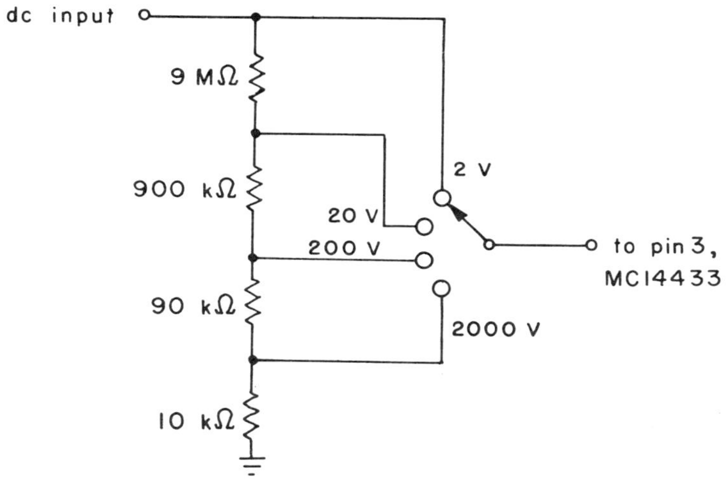

Fig. 8-7. Resistor divider string for increasing full-scale voltage.

At pin 15, the OVERRANGE BLANKING signal, which is normally high, blanks the display when the input voltage exceeds ±2.000 volts. The 75492 inverter/driver is used for interfacing the four display enable outputs to the display during the strobing process.

This basic circuit can be enhanced in several ways. Using the resistor divider string shown in Fig. 8-7, the basic circuit is then capable of measuring full-scale voltages from 1.999 to 1999.9 volts. On the other hand, we are able to measure ac voltages with the op-amp half-wave rectifier circuit of Fig. 8-8. The peak, or dc, output from the rectifier is calibrated to produce the rms reading for a sine wave, so that

$$V_{rms} = 0.707\ V_{peak} \qquad \text{(Eq. 8-1)}$$

A complete digital multimeter with autoranging is fully described in Motorola's Application Note AN-769.*

* Aldridge, D. "Autoranging Digital Multimeter Using the MC14433 CMOS A/D Converter," Motorola Application Note AN-769, 1977.

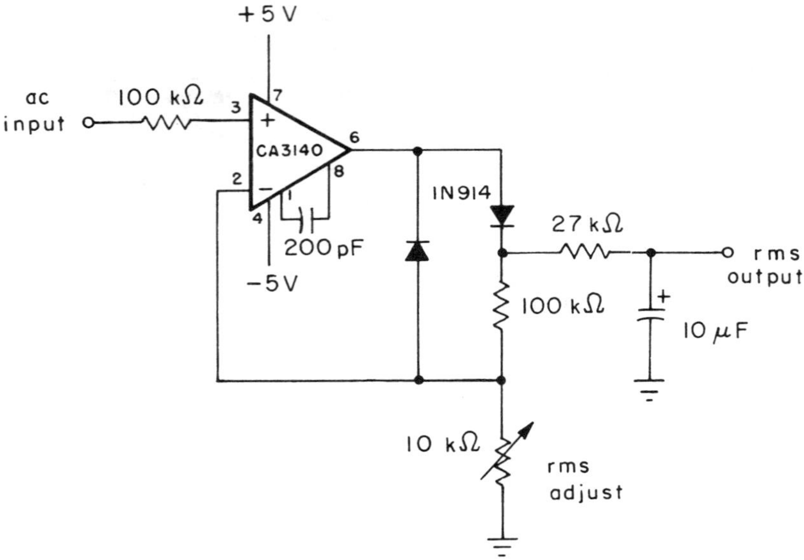

Fig. 8-8. Op-amp half-wave rectifier for measuring rms voltage.

The Intersil ICL7106 and 7107, whose pin diagrams are shown in Fig. 8-9, are both 3-1/2 digit A/D converters similar in function to the MC14433. The 7106 is designed to be interfaced with a liquid crystal display (LCD) and, therefore, includes the necessary backplane drive signal (pin 21). The 7107 will directly drive 0.3 inch or 0.43 inch common anode LED displays.

The general circuits of Figs. 8-10 (7106) and 8-11 (7107) are specified for a 200-mV full scale range. Like the MC14433, both devices can be wired to permit a 2.000 V full scale range by changing the values of the reference resistor R_1, the integrator resistor R_2, and the auto-zero capacitor C_1 to those values listed in Table 8-4.

The MC14433 and ICL7107 devices are designed to operate from ±5 volt supplies. However, when a negative supply is not convenient, as in certain portable applications, a suitable substitute is

Table 8-4. Component Values Required for Full-Scale Voltage

Component	200.0 mV Full Scale	2.000 V Full Scale
R_1	24 kΩ	1.5 kΩ
R_2	47 kΩ	470 kΩ
C_1	0.47 μF	0.047 μF

shown in Fig. 8-12. The input signal is taken from the internal oscillator of either the MC14433 (pin 11) or the 7107 (pin 38). Complete details on digital panel meter design using either the 7106 or the 7107 is described in Intersil's Application Bulletin A023.*

4046 PHASE-LOCKED LOOP

In addition to a voltage-controlled oscillator, as was described in Chapter 5, the 4046 (Fig. 8-13) also contains a phase detector to form a one-chip phase-locked loop when an external low-pass filter is added.

Referring to the block diagram of Fig. 8-14, the 4046 actually has two phase detectors, whose inputs are connected in parallel. The outputs, however, are brought out separately. Phase detector I (sometimes called the *low-noise detector*) is a simple-exclusive-OR type. Consequently, both the input and vco signals must be symmetrical square waves.

Phase detector II (sometimes called the *wideband detector*) is an edge-triggered digital type, which triggers on the positive leading edges of the two inputs. If the input signal, which can be a pulse train having any duty cycle, is lower than the vco (voltage-controlled oscillator) frequency, the output is at logic 0 (V_{SS} or ground). On the other hand, if the input frequency is higher than the vco frequency, the output is at logic 1. If both frequencies are the same, the output of phase detector II is a pulse whose width is proportional to the phase difference. As shown in Fig. 8-15, this output pulse is positive when the vco signal lags the input, and negative when the vco leads the input. One advantage of phase detector II over the other is that the wideband detector is insensitive to harmonic multiples of the input frequency, while the exclusive-OR type may lock onto these harmonic multiples.

Also shown in the block diagram is an uncommitted 5.4-volt zener diode, which can be used if regulation of the power supply voltage is necessary. The frequency of the vco is a minimum when the input control voltage of the vco is zero, and increases linearly to a maximum when the control voltage equals the supply voltage. As shown in the basic circuit of Fig. 8-16, the minimum-maximum vco range is set by the external components R_1 and C_1, while the minimum vco frequency is controlled by R_2 and C_1. Momentarily connecting pin 9 to ground consequently causes the vco to oscillate at its minimum frequency, while connecting pin 9 to the supply voltage gives the maximum frequency. If for some applications it is not necessary

* Fullagar, D., and Dufort, M. "Low Cost Digital Panel Meter Designs," Intersil Application Bulletin A023, 1977.

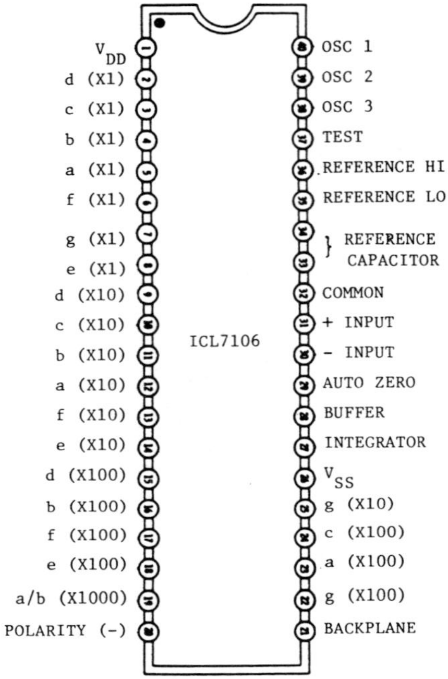

(A) ICL7106 pin diagram (LCD output).

Fig. 8-9. Intersil

to have the vco on at certain times, the vco can be turned off by tying the INHIBIT input (pin 5) to the supply voltage, which will also minimize power consumption. Otherwise, the INHIBIT input is connected to V_{SS} or ground.

To get the 4046 or any other phase-locked loop system to work, a loop filter is required between the output of the phase detector and the vco input. Although the loop filter can take on several forms, the *lag-lead* circuit of Fig. 8-17 is frequently used. The ratio of R_1/R_2 basically determines the damping of the loop, while the R_1C time constant determines the settling time. Generally, the value of R_1 is made anywhere from 3 to 10 times larger than R_2. Specific details on loop filter design can be found in either *Design of Phase-Locked Loop Circuits, with Experiments*, or *Phaselock Techniques.**

The PHASE PULSES output (pin 1) of phase detector II can be NORed with the output of phase detector I (pin 2) to form a lock detector, as shown in Fig. 8-18. The output of the last NOR gate will be at logic 1 when the loop is locked.

---

* Gardner, F. M. *Phaselock Techniques*. John Wiley & Sons, Inc., New York, 1966.

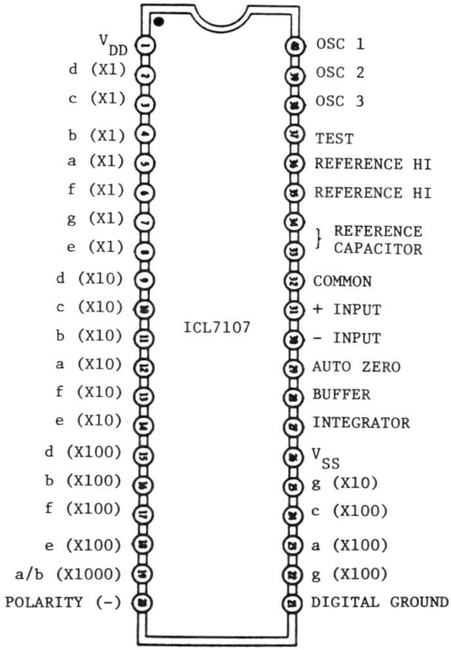

(B) ICL7107 pin diagram (LED output).

A/D converters.

By breaking the feedback path of the loop (pins 4 and 3) and inserting a divide-by-*N* counter between the vco and the phase detector, the 4046 can be used as a frequency synthesizer or multiplier. As an example, the resolution of a typical frequency counter can be improved from ±1 Hz to ±0.01 Hz for low frequency signals by using a pair of decade counters (4017, 74C90, etc.) in cascade, as shown in Fig. 8-19. Thus, an input frequency of 52.83 Hz will be displayed as 5283 Hz on the counter. Otherwise, all you would observe is either 52 or 53 Hz. By making the modulus of the counter chain equal to 60, this same technique is used to display the input frequency in terms of *cycles per minute.* This is extremely useful when measuring the frequency of physiological events such as respiration (breaths per minute), or heart rate (beats per minute).

A TWO-CHIP FREQUENCY COUNTER

With the ICM7207 oscillator controller and the ICM7208 7-decade counter-decoder-driver, we are able to build a battery-powered counter capable of both frequency (up to 2 MHz) and period measurements.

Shown in Fig. 8-20A, the ICM7207 is a crystal-controlled oscillator having outputs for count, latch, reset, and display multiplex functions, for the 7208. When a 6.5536-MHz crystal is used, it is divided by 2^{12} to produce the 1600-Hz signal that multiplexes the display of the counter. The 1600-Hz signal is divided further to provide counting gate signals of either 0.01 second or 0.1 second. The timing diagram for the four output signals is shown in Fig. 8-20B.

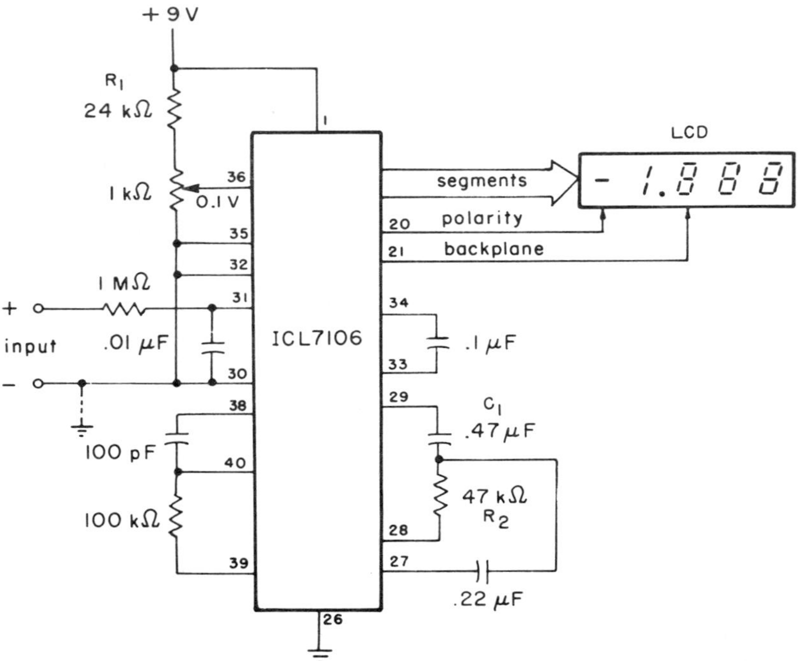

Fig. 8-10. 7106-LCD voltmeter.

The ICM7208, shown in Fig. 8-21, provides the following functions: a 7-decade counter and decoder/driver, multiplexer, plus logic for display reset, blanking, input inhibit, and display on/off. The complete two-chip counter is shown in Fig. 8-22.

Although the upper range of the counter is typically 2 MHz, this range can be increased to 20 or 200 MHz by the addition of either a divide-by-10 or a divide-by-100 counter at the input of the counter. On the other hand, the resolution of the counter circuit can be increased with the addition of a 4046 phase-locked loop multiplier, as described in the previous section. By interchanging the roles of the COUNTER (pin 12) and INHIBIT (pin 13) inputs, the circuit

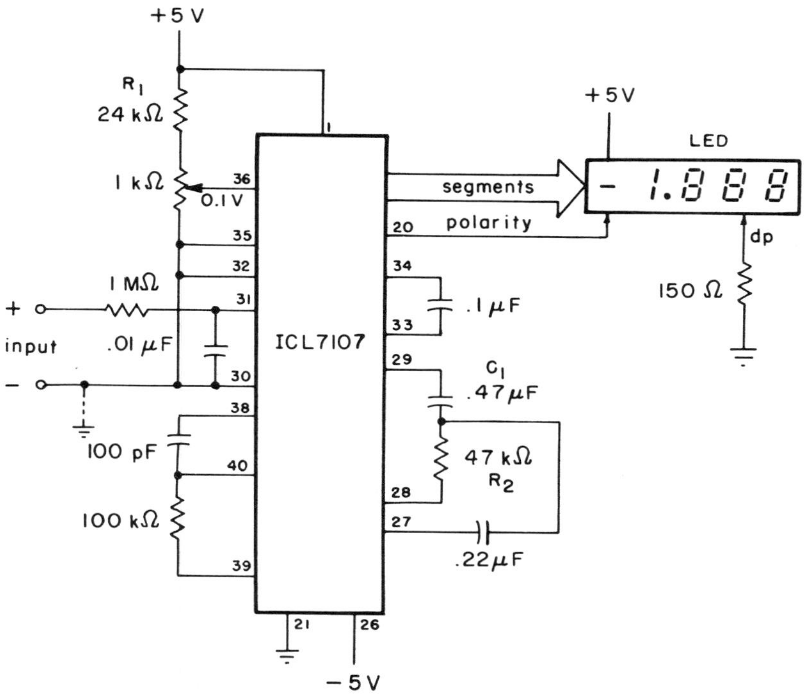

Fig. 8-11. 7107-LED voltmeter.

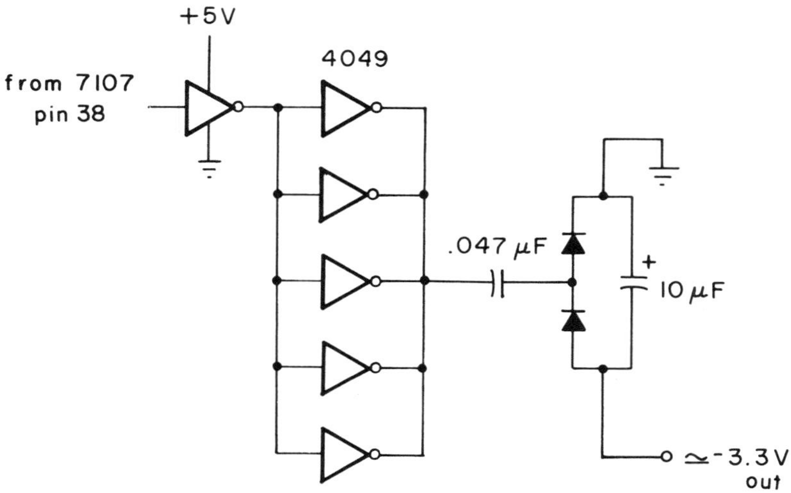

Fig. 8-12. Negative supply voltage from a +5-volt supply.

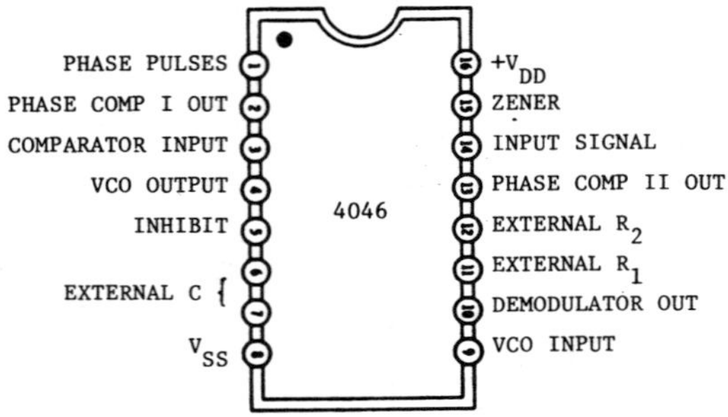

Fig. 8-13. 4046 phase-locked loop pin diagram.

is then able to measure time periods rather than frequency. Complete design and construction details for this frequency counter is described in Intersil's Application Bulletin A015.*

74C926 4-DIGIT COUNTER

National Semiconductor's 74C926 4-digit counter, as shown by the pin diagram of Fig. 8-23, is a self-contained 4-decade counter with multiplexed display. As shown by the block diagram of Fig. 8-24,

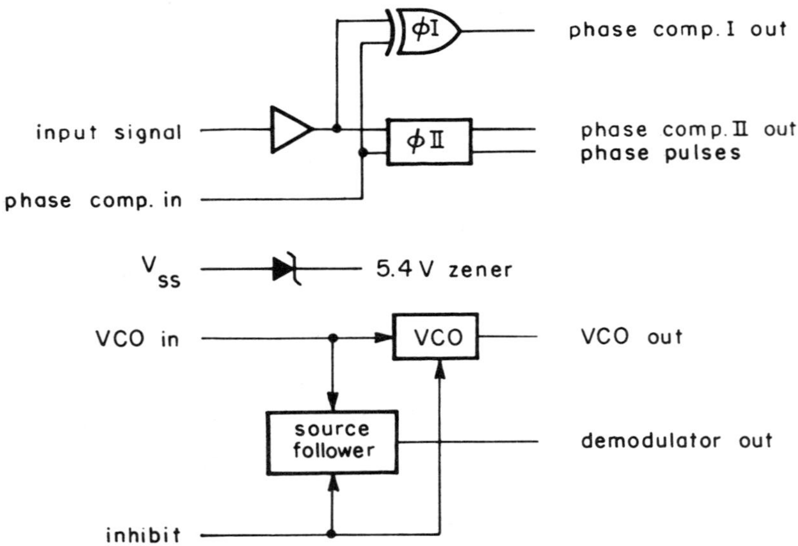

Fig. 8-14. Block diagram.

* "Design for a Battery Operated Frequency Counter," Intersil Application Bulletin A015, 1977.

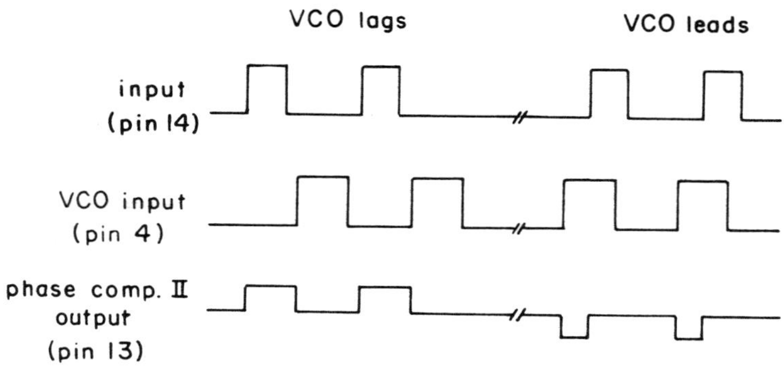

Fig. 8-15. 4046 waveforms.

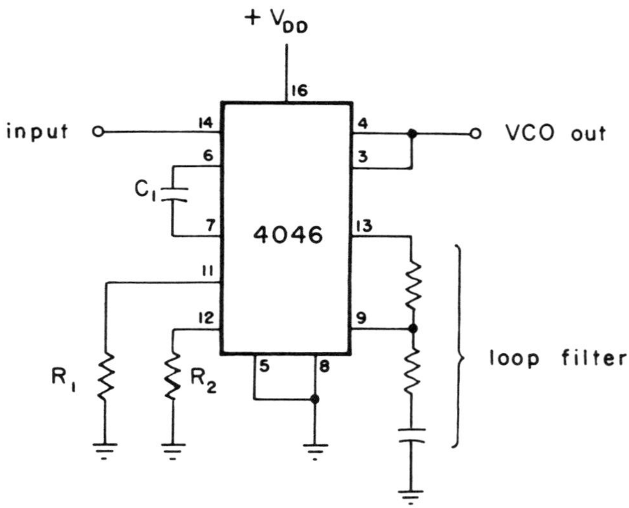

Fig. 8-16. 4046 phase-locked loop circuit.

Fig. 8-17. Lag-lead low-pass loop filter.

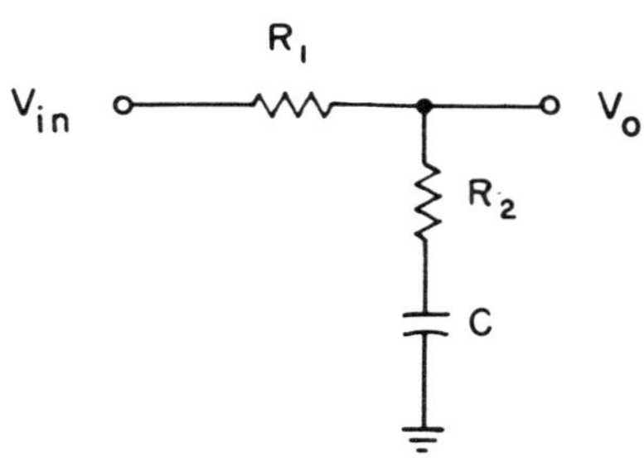

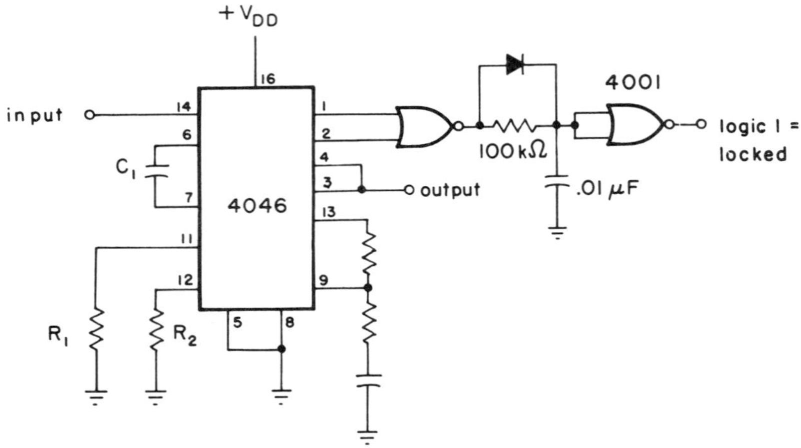

Fig. 8-18. NOR gate lock detector.

the 74C926 internally contains an output latch, npn output sourcing drivers for 7-segment LED displays, and a multiplexer, which has its own free-running oscillator.

Using the basic operating circuit of Fig. 8-25, the counters count up on the negative edge of each input clock pulse. To reset the counters, the RESET input (pin 13) is momentarily brought high. The CARRY OUT signal at pin 14 is used for cascading two or more 74C926 devices. It goes high when the count reaches 6000 and then

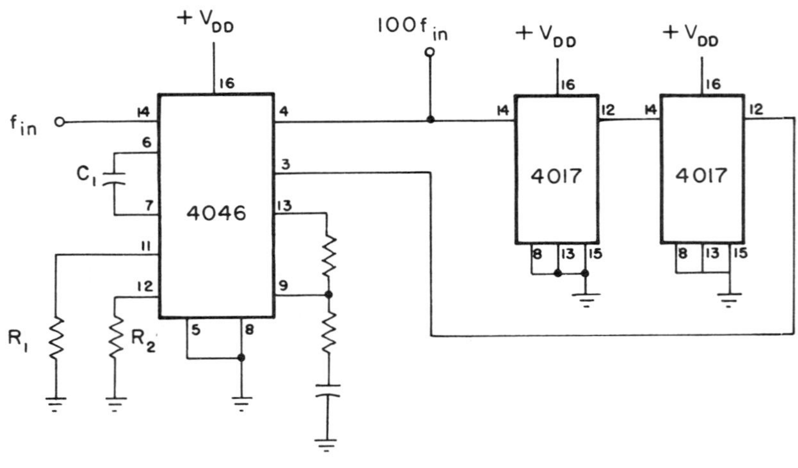

Fig. 8-19. X100 frequency synthesizer or prescaler.

low at a count of 0000. A low signal at the LATCH ENABLE input will latch the present count to the output latches. When the DISPLAY SELECT input is low, the display will read the count that is held in the four 4-bit latches. Otherwise, the display will read the count that is presently in the counters. Fig. 8-26 shows the typical switching time waveforms required.

(A) Pin diagram.

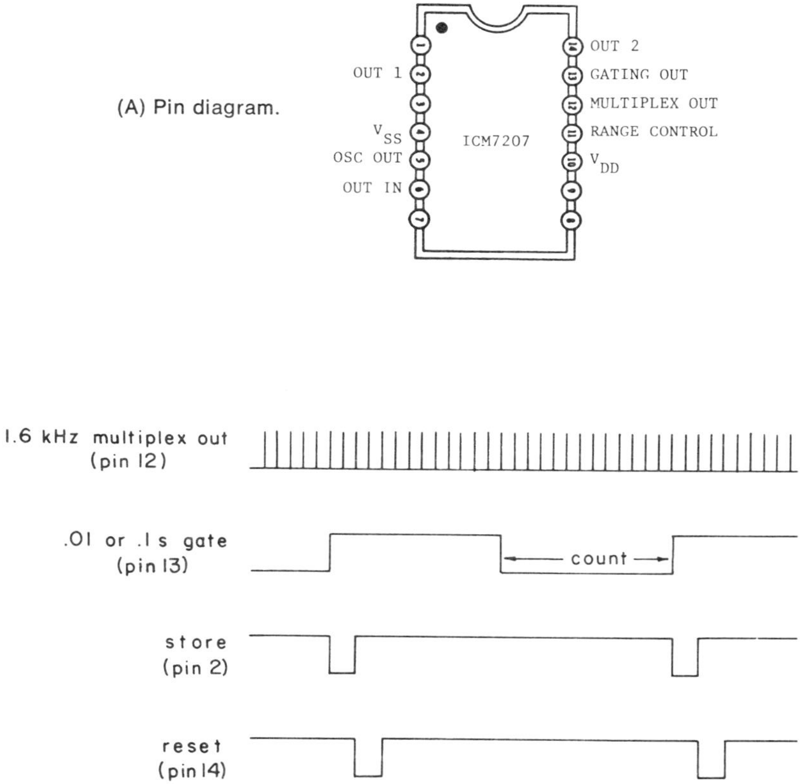

(B) Output waveforms.

Fig. 8-20. ICM7207 crystal-controlled oscillator.

AN INTRODUCTION TO THE EXPERIMENTS

As an example of the use and operation of the specialized CMOS devices described in this chapter, the following experiments will use the 4046 CMOS phase-locked loop as a representative example.

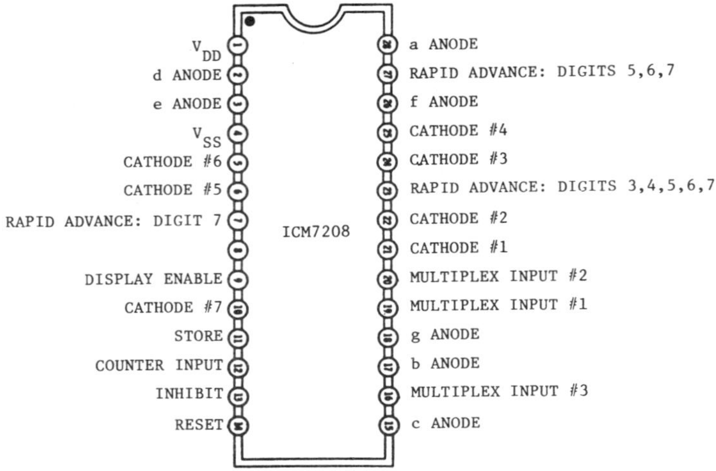

Fig. 8-21. ICM7208 7-decade counter and decoder/driver.

The following experiments that you will perform can be summarized as follows:

Experiment No.	*Purpose*
1	Demonstrates the operation of the 4046 phase-locked loop.
2	Demonstrates the operation of a "loss-of-lock" indicator with the 4046 phase-locked loop.
3	Demonstrates the operation of a simple frequency synthesizer that can be used as a multiplying prescaler for frequency counters.

EXPERIMENT 1

Purpose

The purpose of this experiment is to demonstrate the operation of the 4046 monolithic CMOS phase-locked loop.

Step 1

Set your oscilloscope for the following settings:

- Channel 1: 0.5 V/division
- Time base: 0.5 ms/division

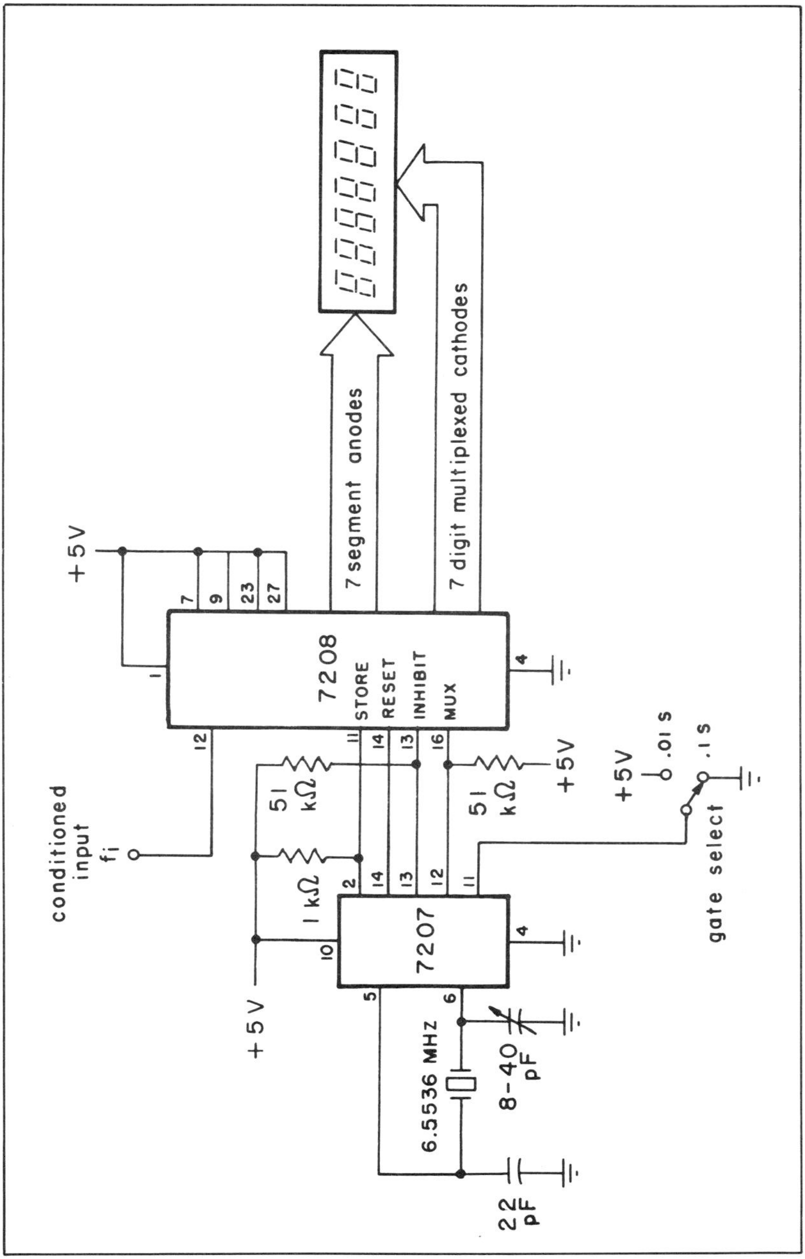

Fig. 8-22. Two-chip frequency counter.

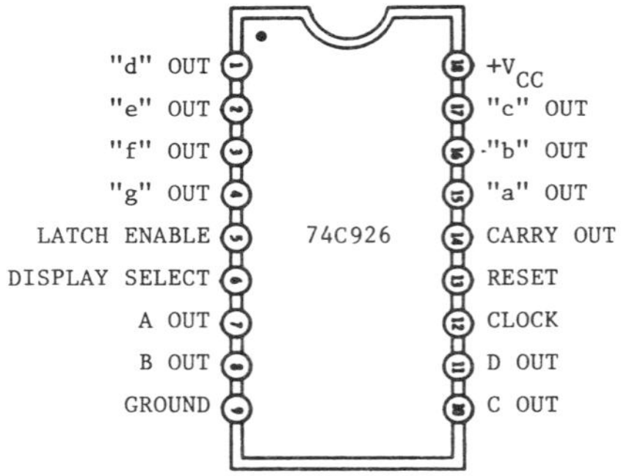

Fig. 8-23. 74C926 pin diagram.

Step 2

Wire the circuit as shown in the schematic diagram and apply power to the breadboard. Adjust the output of the function generator (sine wave) at approximately 1 kHz with the frequency counter, and the peak-to-peak voltage at 2 volts (i.e., 4 vertical divisions). Now connect the frequency counter to pins 3 and 4 of the 4046 device. What do you notice about the output frequency of the phase-locked loop?

The output frequency of the phase-locked loop should be the same as its input.

Step 3

With an extra piece of wire, connect pin 9 of the 4046 to *ground*. Record the resultant output frequency of the phase-locked loop:

$$f_L = \_\_\_\_\_\_\_\_\_\_ \text{Hz}$$

This output frequency is the lower limit of the vco, which is determined by the 0.1 μF capacitor connected between pins 6 and 7 and the 100 kΩ resistor connected between pin 12 and ground.

Step 4

Now, with the same wire, connect pin 9 to the +5-volt supply. You should observe an output frequency that is higher than the one you measured in Step 3. Record this frequency:

$$f_H = \_\_\_\_\_\_\_\_\_\_ \text{Hz}$$

This output frequency is the upper limit of the vco, which is basically determined by the 560 Ω resistor and the 0.1 μF capacitor between pins 6 and 7.

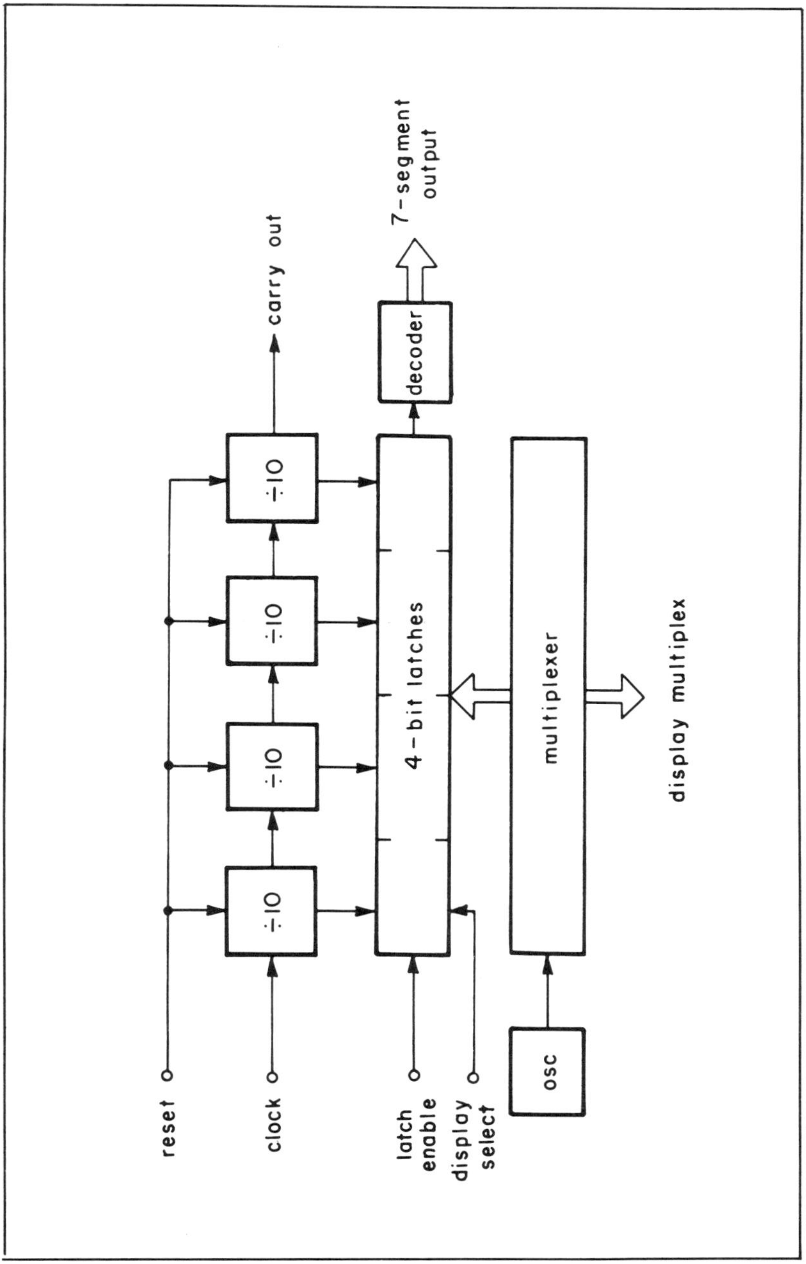

Fig. 8-24. Block diagram.

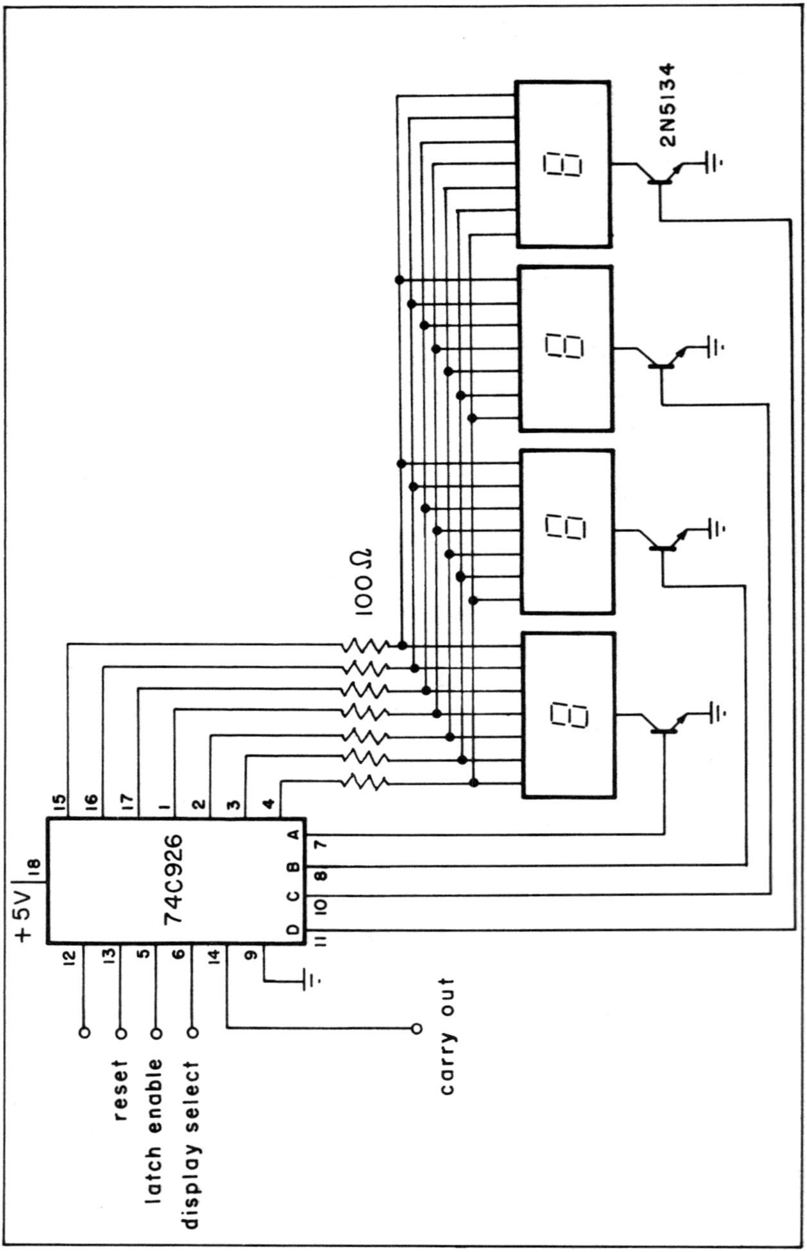

Fig. 8-25. 4-decade counter.

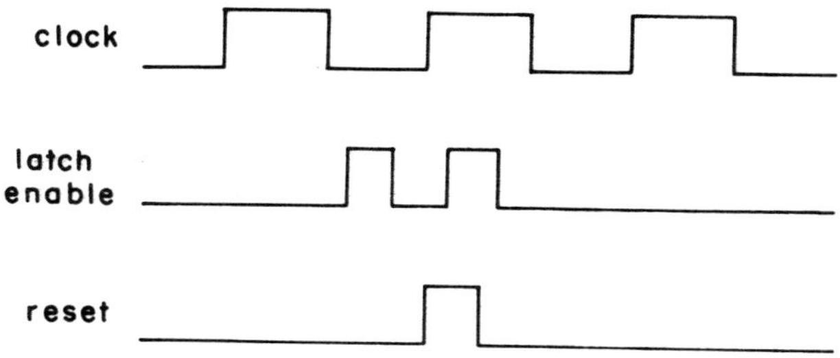

Fig. 8-26. Input switching waveforms.

Pin Configuration of Integrated Circuit Chip (Fig. 8-27)

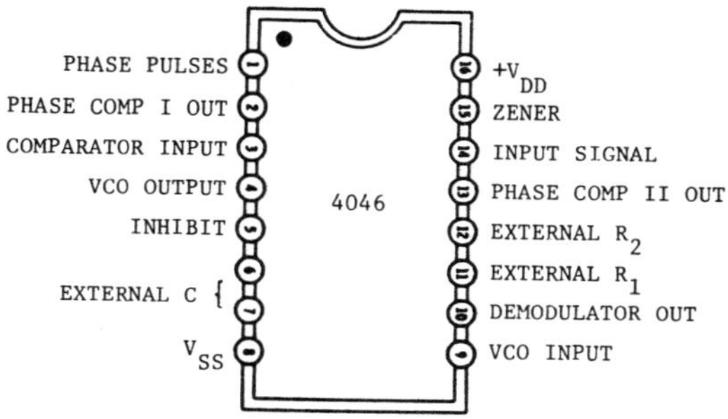

Fig. 8-27.

Schematic Diagram of Circuit (Fig. 8-28)

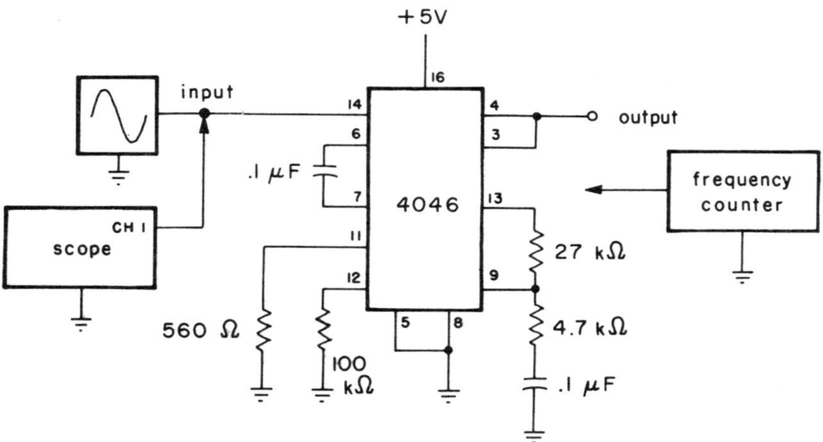

Fig. 8-28.

Step 5

Now remove this connection between pin 9 and the +5-volt supply. You should again measure an output frequency that is the same as the frequency of the function generator as in Step 2.

Step 6

Connect the frequency counter to pins 3 and 4 of the 4046 integrated circuit. Now slowly increase the frequency of the function generator. What do you observe on the frequency counter?

You should observe that the output frequency also increases! In fact, the output frequency follows the changes of the input frequency and should be exactly equal. Check the input frequency to confirm this statement.

Step 7

While watching the output frequency of the phase-locked loop, continue to slowly increase the input frequency and stop when the output frequency does not increase any further. Measure the input frequency and record your result:

$$f_{in}(H) = \_\_\_\_\_\_\_\_ \text{ Hz}$$

You should find that this frequency is about the same as the frequency which you measured in Step 4, the *upper limit of the vco.* The phase-locked loop then follows input frequency changes for frequencies below this upper limit.

Step 8

Now decrease the input frequency while observing the frequency counter. At some point the output frequency will remain constant. Measure the input and record your result:

$$f_{in}(L) = \_\_\_\_\_\_\_\_ \text{ Hz}$$

You should find that this frequency is about the same as the frequency which you measured in Step 3, the *lower limit of the vco.* Consequently, the phase-locked loop circuit follows changes in the input frequency for any frequency between the lower and upper limits of the vco. Therefore, the loop is *locked.* The range over which the phase-locked loop follows changes in the input frequency is called the *lock range.* To determine the lock range from your measurements, subtract the value you measured in Step 8 from the value of Step 7, and record your result:

lock range = ________ Hz

The lock range can be changed by simply changing the value of the resistor connected to pin 11 or 12. Decreasing the 100 kΩ resistor at pin 12, for example, *increases* the *lower* frequency range.

Keep this circuit on your breadboard, as it will be used in the next experiment.

EXPERIMENT 2

Purpose

The purpose of this experiment is to demonstrate a "loss-of-lock" indicator with the phase-locked loop of Experiment 1.

Pin Configuration of Integrated Circuit Chip (Fig. 8-29)

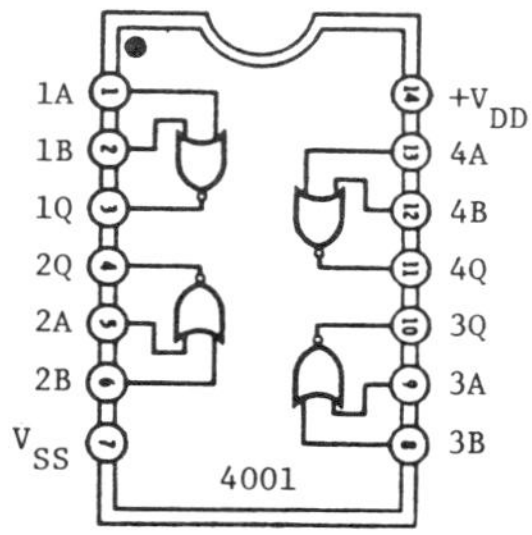

Fig. 8-29.

Schematic Diagram of Circuit (Fig. 8-30)

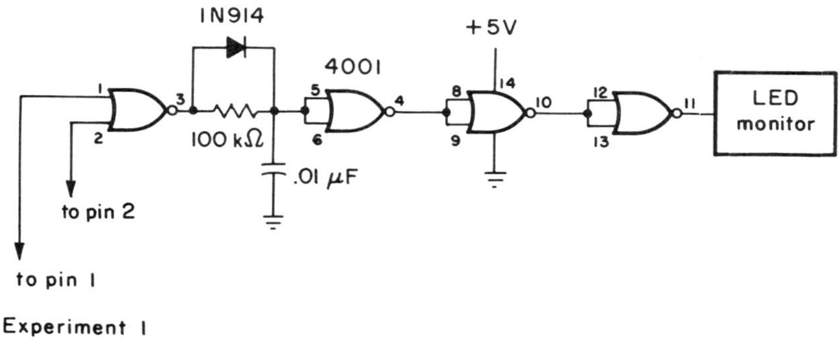

Fig. 8-30.

Step 1

Wire the loss-of-lock circuit as shown in the schematic diagram. Connect pin 1 of the 4001 CMOS NOR gate to pin 1 of the 4046 phase-locked loop (PHASE PULSES output of comparator II), and pin 2 of the 4001 NOR gate to pin 2 of the 4046 (phase comparator I output). Make sure that you have connected the 1N914 diode correctly across the 100 kΩ resistor.

Step 2

Apply power to the breadboard and set the input frequency at approximately 500 Hz. Is the LED monitor lit or unlit?

The LED monitor should be lit since the 500-Hz input frequency is within the lock range of the loop circuit, which you determined in Steps 7 and 8 of the previous experiment. When the loop is phase-locked, the output of the loss-of-lock circuit is at logic 1.

Step 3

Increase the input frequency just past the upper limit of the VCO (Step 7 of the previous experiment). What happens to the LED monitor?

The LED monitor should now be unlit, indicating that the phase-locked loop is now unlocked, since the input frequency is now outside the lock range of the loop. In some cases, we have noticed that the LED monitor flickers on and off several times as the loop becomes unlocked. This behavior is due to the transient behavior of the loop.

Step 4

Change the input frequency to 1 kHz. From the LED monitor, is the loop locked or unlocked?

The LED monitor should be lit since the 1-kHz input signal is within the lock range of the loop. We can then use such a circuit with the 4046 phase-locked loop to visually indicate whether the loop is locked or not.

EXPERIMENT 3

Purpose

The purpose of this experiment is to demonstrate the operation of a 4046 phase-locked loop and a 4017 decade counter as an X10 frequency multiplier or prescaler.

Pin Configuration of Integrated Circuit Chips (Fig. 8-31)

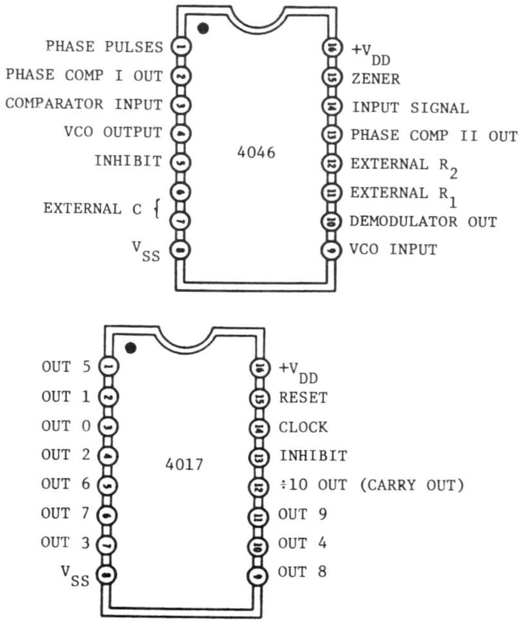

Fig. 8-31.

Step 1

Set your oscilloscope for the following settings:

- Channel 1: 0.5 V/division
- Time base: 10 ms/division
- AC coupling

Step 2

Wire the circuit as shown in the schematic diagram and apply power to the breadboard. Connect the frequency counter to pin 14 of the 4046 phase-locked loop. Adjust the input frequency so that

Schematic Diagram of Circuit (Fig. 8-32)

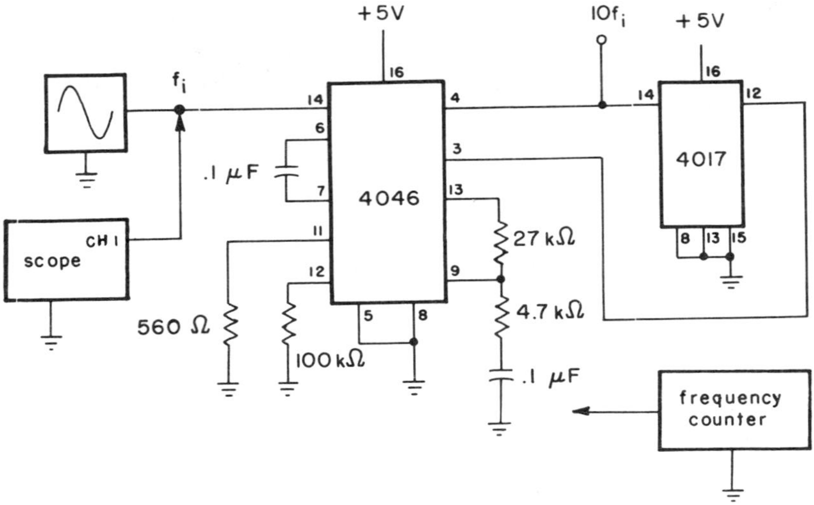

Fig. 8-32.

the input frequency (f_i) is somewhere between 80 and 90 Hz. In addition, adjust the peak-to-peak input sine wave at 2 volts.

Step 3

Measure the input frequency and record your result:

f_i = ________ Hz

Step 4

Now connect the frequency counter to pin 4 of the 4046 device. Measure the output frequency and record your result:

f_o = ________ Hz

What relationship do you notice between the frequency that you measured in this step and the one in Step 3?

The output frequency should be 10 times larger than the input. The input frequency that you measured in step 3 normally has a resolution of ±1 Hz. By using this circuit to multiply the input frequency by 10, we are then able to measure the input frequency with a resolution of ±0.1 Hz. As an example, if you measured an input frequency of 87 Hz in Step 3, this means that the input frequency could range from 86 to 88 Hz. If the measured output frequency was 867 Hz, then the input frequency is 86.7 Hz, not 87 Hz as was presumably measured in Step 3. The resolution of the frequency counter has now been increased by one significant digit.

Step 5

Choose any input frequency between 20 Hz and 300 Hz and then measure both the input and output frequencies of the synthesizer circuit. Over this input range you should be convinced that it is now possible to measure an input frequency with a resolution of ±0.1 Hz instead of ±1 Hz.

APPENDIX A

TTL to CMOS Conversion Chart

As an aid in selecting a given CMOS device for those having experience with TTL devices, the following chart summarizes the many CMOS devices that are functional equivalents of their TTL counterparts. CMOS devices are listed primarily using the 4000 series numbering system introduced by RCA. Other manufacturers generally follow the same numbering system, but with different prefixes. Motorola's CMOS devices belong to the MC14000 series, while those manufactured by Fairchild belong to the 34000 series. Consequently, a 4081 (RCA) is equivalent to both an MC14081 (Motorola) and a 34081 (Fairchild). Those devices listed as having the 74C00 series are pin-for-pin equivalents of the 7400 series of TTL devices.

Table A-1. TTL to CMOS Conversion Chart

TTL	CMOS	TTL	CMOS
7400	4011, 74C00	74107	4027, 74C107
7401	40107	74110	4095
7402	4001, 74C02	74111	4027
7404	4009, 4049, 74C04	74114	40106
7406	4009, 4049	74121	4047, 4098
7407	4010, 4050	74122	4047, 4098
7408	4081, 74C08	74123	4098, 74C221
7410	4023, 74C10	74125	4502
7414	MC14584, 74C14, 40106	74126	4502
7420	4012, 74C20	74132	4093
7425	4002	74136	4030, 4070
7427	4025	74141	4028
7428	4001	74145	4028
7430	4068, 74C30	74148	4532
7432	4071, 74C32	74150	4067
7437	4011	74151	4051, 4097, 74C151
7440	4012	74152	4051, 4097
7442	4028, 74C42	74153	4052
7445	4028	74154	4514, 4515, 74C154
7446	4055, 4511	74155	4555, 4556
7447	4055, 4511	74156	4555, 4556
7448	4511, 74C48	74157	4019, 74C157
7449	4511	74160	74C160
7450	4085	74161	74C161
7453	4086	74162	74C162
7454	4086	74163	74C163
7470	4096	74164	4015, 74C164
7472	4095	74165	4021, 74C165
7473	4027, 74C73	74166	4014
7474	4013, 74C74	74167	4527
7475	4042	74173	4076, 74C173
7476	4027, 74C76	74174	74C174
7477	4042	74175	74C175
7478	4027	74178	4035
7483	4008, 74C83	74179	4035
7485	4063, 74C85	74180	40101
7486	4030, 4070, 74C86	74181	40181
7489	74C89	74182	40182
7490	4518, 74C90	74190	4510
7491	4015, 4094	74191	4516
7493	4520, 74C93	74192	40192, 74C192
7494	4035	74193	40193, 74C193
7495	40104, 40194, 74C95	74194	40104, 40194
7499	40104, 40194	74195	4035, 40195, 74C195
74100	4034	74198	4034
74104	4095	74200	4061, 74C200
74105	4095		

APPENDIX B

Breadboarding Aids

In addition to breadboarding sockets, there are a number of useful commercial breadboarding aids available. Among these are an extensive line of aids called OUTBOARDS®.* This appendix describes the OUTBOARDS that are useful in performing many of the experiments described in this book and that were illustrated by

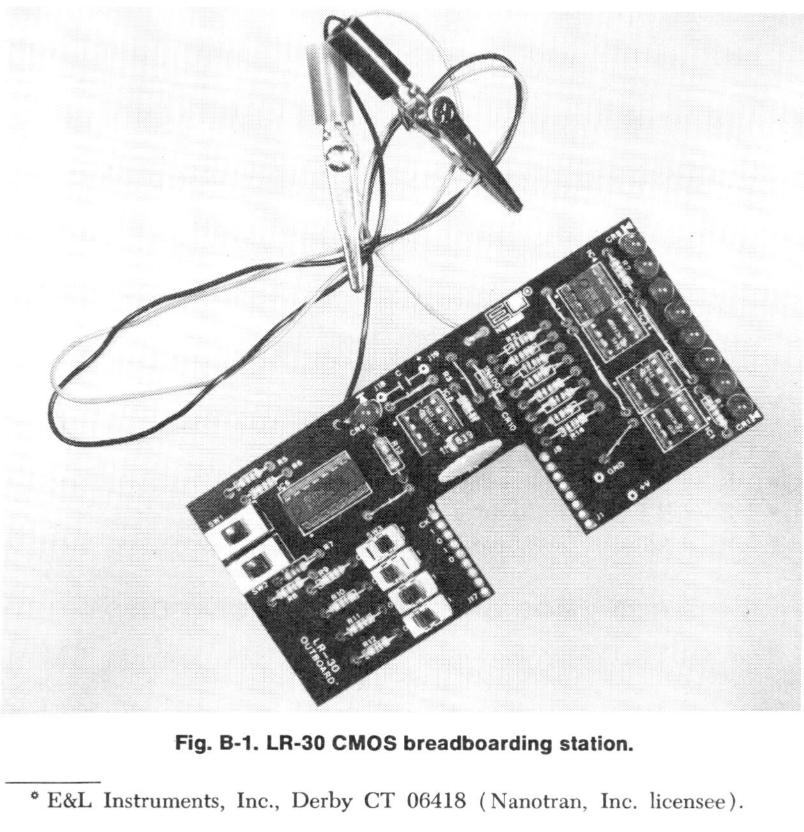

Fig. B-1. LR-30 CMOS breadboarding station.

* E&L Instruments, Inc., Derby CT 06418 (Nanotran, Inc. licensee).

Fig. B-2. LR-31 function generator OUTBOARD.

schematic diagrams in Chapter 2. Each OUTBOARD attaches directly to one of the popular breadboarding sockets, making contact with the necessary power supply connections. The following OUTBOARDS are described:

- LR-30 CMOS Breadboarding Station
- LR-31 Function Generator OUTBOARD
- LR-32 4-Decade Counter OUTBOARD
- LR-33 Quartz Crystal OUTBOARD

LR-30 CMOS BREADBOARDING STATION

This OUTBOARD, shown in Fig. B-1, is a complete CMOS breadboarding station. It contains two debounced logic switches, or pulsers, 4 logic switches, 8 LED monitors, and a variable frequency square wave clock. In addition, the circuitry is protected from incorrect power supply connections. The LR-30 operates over a range from 3 to 15 volts.

LR-31 FUNCTION GENERATOR OUTBOARD

This OUTBOARD, shown in Fig. B-2, uses an XR-2206 integrated circuit to generate sine, square, and triangle waveforms with adjustable frequency, amplitude, and dc offset. Frequencies of 0.01 Hz to 1 MHz are possible, depending on the value of an external

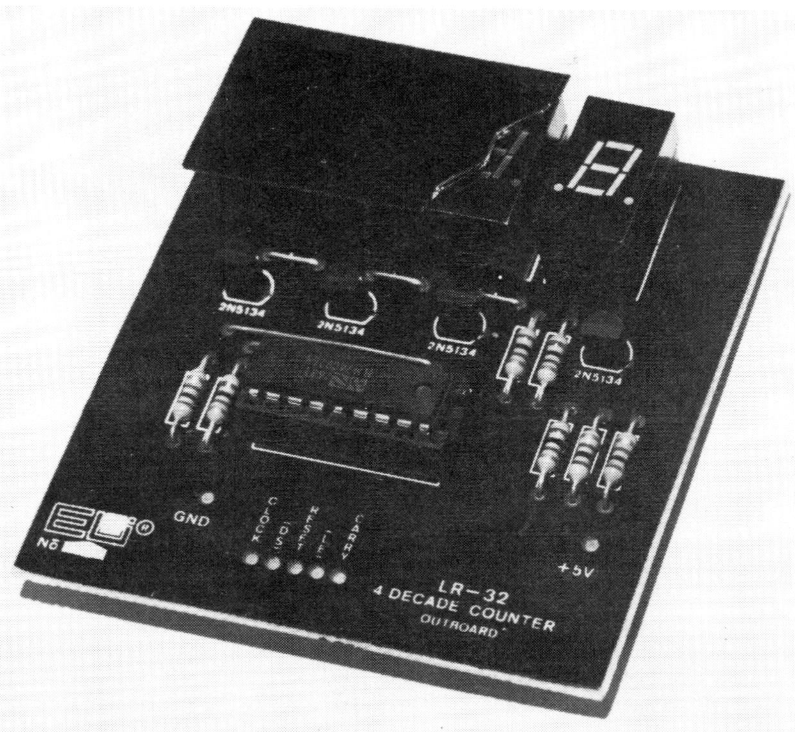

Fig. B-3. LR-32 4-decade counter OUTBOARD.

capacitor. With a single capacitor value, the dynamic range is greater than 1000:1.

LR-32 4-DECADE COUNTER OUTBOARD

Using a 74C926 integrated circuit, the LR-32, shown in Fig. B-3, is a complete 4-decade counter including four decade counters, latches, display drivers, display multiplexing circuitry, and four 7-segment LED displays. The CARRY OUTPUT of the LR-32 is available so that two or more LR-32s can be easily cascaded.

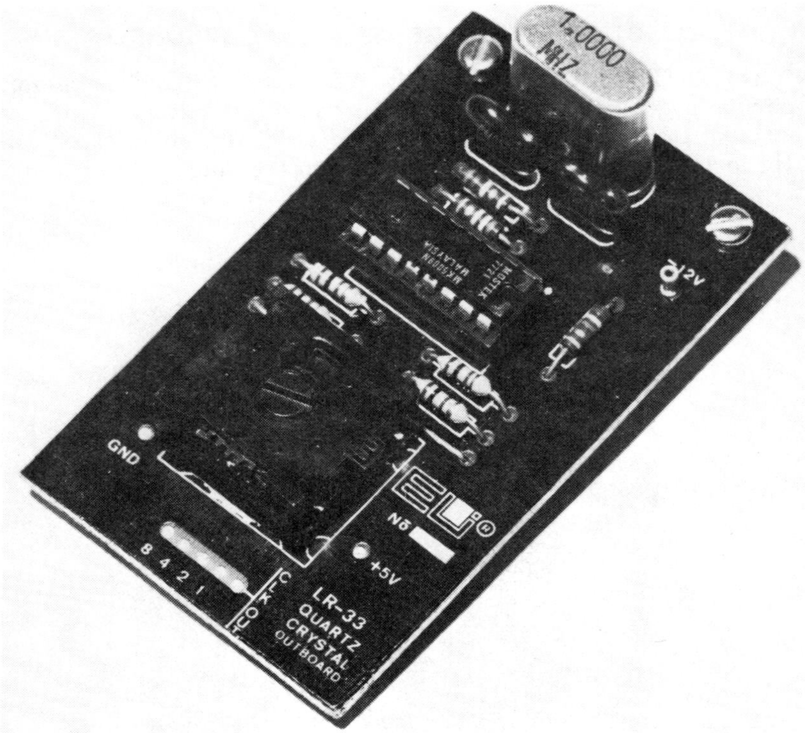

Fig. B-4. LR-33 quartz crystal OUTBOARD.

LR-33 QUARTZ CRYSTAL OUTBOARD

This OUTBOARD, shown in Fig. B-4, is similar to the schematic diagram shown in Chapter 2 (Fig. 2-10). With the on-board thumbwheel switch, you can select a crystal-controlled output frequency from 1 MHz to 0.01 Hz in factors of 10.

Index

D

E

F

G

H

I

J

L

M

TO THE READER

This book is one of an expanding series of books that will cover the field of basic electronics and digital electronics from basic gates and flip-flops through microcomputers and digital telecommunications. We are attempting to develop a mailing list of individuals who would like to receive information on the series. We would be delighted to add your name to it if you would fill in the information below and mail this sheet to us. Thanks.

1. I have the following books:

____

____

____

____

____

____

____

____

____

____

____

____

2. My occupation is: ☐ student ☐ teacher, instructor ☐ hobbyist

☐ housewife ☐ scientist, engineer, doctor, etc. ☐ businessman

☐ Other: ______________________________

Name (print): ______________________________

Address ______________________________

City ______________________________ State ______________

Zip Code ______________________________

Mail to:

Books
P.O. Box 715
Blacksburg, Virginia 24060